Animal Genetics for Chemists

This book features humans (a lot), other mammals (a good deal) and, occasionally, other animals to illustrate principles.

Animal Genetics for Chemists

Ralph G. Wilkins
New Mexico State University, USA

THE QUEEN'S AWARDS
FOR ENTERPRISE:
INTERNATIONAL TRADE
2013

Print ISBN: 978-1-78262-760-9

A catalogue record for this book is available from the British Library

The Royal Society of Chemistry is a charity, registered in England and Wales, Number 207890, and a company incorporated in England by Royal Charter (Registered No. RC000524), registered office: Burlington House, Piccadilly, London W1J 0BA, UK, Telephone: +44 (0) 207 4378 6556.

Visit our website at www.rsc.org/books

Printed in the United Kingdom by CPI Group (UK) Ltd, Croydon, CR0 4YY, UK

Preface

Genetics and chemistry are intertwined disciplines. DNA, the workhorse of genetics, is after all a chemical that undergoes chemical reactions with which the chemist and biochemist are quite familiar. Five of the last ten Nobel Prizes in Chemistry were related to genetics. I decided a few years ago to study genetics without any background in the subject. Later, I was bold and I hope not foolish to write this short book by a chemist for chemists and hopefully also for a broad readership who are interested in genetics but do not require the detailed information found in larger and more authoritative text books. Consequently, this book only scratches the surface of, or even excludes, many aspects of genetics. I hope, however, that enough of the basics have been included to enable the reader to understand the number of news items, reporting small and big breakthroughs in genetics and their implications to society, that are pouring out almost daily from scientific journals, magazines and newspapers.

I decided to confine the book to animal genetics to limit its size but also because many of the concepts first established with bacteria and plants can be explained using human and animal behavior. Humans are treated separately from other mammals and both form a good part of the book, although non-mammals have a place sometimes in illustrating principles. Believing in the saying that a picture is worth a thousand words, I have

Animal Genetics for Chemists
By Ralph G. Wilkins
© Ralph G. Wilkins 2017
Published by the Royal Society of Chemistry, www.rsc.org

supported the text with nearly 100 figures. Ideas have been supported, where possible, with examples using human and animal disorders and diseases, of which I have used many.

The author is grateful to a number of people who have been very helpful in making the book a reality. Jon Griffith and Robert Walters read the book at an early stage and made useful corrections. The Wednesday group were always interested and encouraging. Many thanks are due to the people at the Royal Society of Chemistry (Sylvia Pegg, Catriona Clarke and Katie Morrey) for helpful advice. My special thanks go to Janet Freshwater who was especially helpful and patient throughout the various, sometimes frustrating, stages between submission and acceptance of the book by the Royal Society of Chemistry.

My thanks go to a number of reviewers for helpful suggestions. In particular, I am greatly indebted to Doug Vernimmen (Roslin Institute, University of Edinburgh) for a very diligent reading of the book in which he detected a number of errors and made very many helpful suggestions for possible additions. Any errors or shortcomings that remain are, of course, of my own making.

Finally, the book would have not been possible without the considerable help and support from my wife, Pat. She was courageous to transcribe all the nearly 100 figures from my rough sketches. She helped considerably with my computer illiteracy and with all the things that are necessary in the construction of a book.

All these people have my heartfelt thanks!

Ralph G. Wilkins
Emeritus Professor of Chemistry
New Mexico State University
Las Cruces, New Mexico, USA

This book is dedicated to my wife Pat – my one and only love.

Contents

Animal Genetics for Chemists
By Ralph G. Wilkins
© Ralph G. Wilkins 2017
Published by the Royal Society of Chemistry, www.rsc.org

CHAPTER 1

The Material of Genetics

1.1 THE CELL

In the very beginning is the single cell within a porous membrane. In most developed animals, the single fertilized egg cell
eventually becomes very many cells by a remarkably faithful
reproduction in a process called mitosis. It has been estimated
(see Section 1.1.3) that the number of cells making up an average
human body is in the range of 10 trillion to 100 trillion
$(10–100\times10^{12})$ and with about the same number of bacterial
cells. The large number of human cells will be a mixture of
various irregular shapes and sizes often related to the manner in
which the cells individually support nerve transmission, muscle
contraction, epidermal function and so on in the working organism. These different cell types exist in animals as diverse as
fruit flies, roundworms, guinea pigs, mice and humans, all of
which have played and will play important roles in our understanding of the contribution of genetics to life. In mammals,
there are over 200 different types of cells, ranging in size from
one of the largest, namely the female egg, which is just visible to
the naked eye, to one of the smallest, the sperm cell, which is
produced in the millions. The human egg is round and about
150–200 micrometers (μm) in diameter. Fairly immobile, it is
formed before birth and remains until menopause in the female.
The human sperm is linear-like, with a tail about 50 μm long and

Animal Genetics for Chemists
By Ralph G. Wilkins
© Ralph G. Wilkins 2017
Published by the Royal Society of Chemistry, www.rsc.org

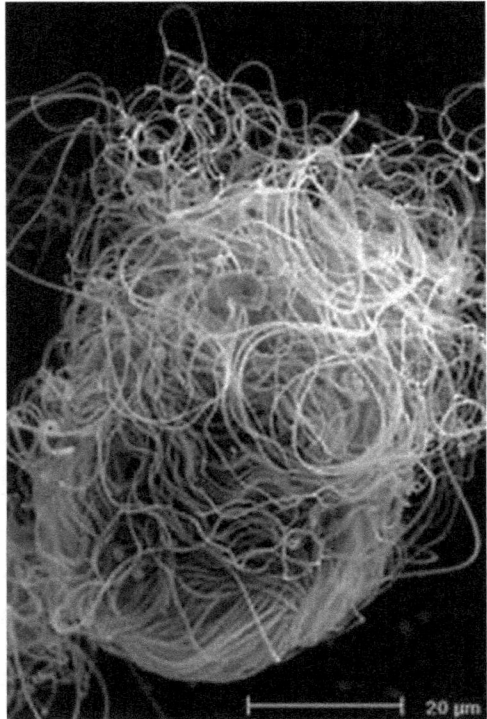

Figure 1.1   SEM showing a coiled ball of fruit fly sperm. Most fruit flies are
under 15 mm long. The fruit fly *Drosophila bifurca* produces sperm
more than 5 cm in length as a coiled ball. This is 1000 times longer
than that of the human.
Reprinted by permission from Macmillan Publishers Ltd: A. Bjork
and S. Pitnick, *Nature*, 2006, **441**, 7094, copyright 2006.

4 µm wide. It is very mobile, using its tail for propulsion. It is
formed during puberty and remains often until death in the
male. These reproductive cells differ widely within the animal
kingdom. In general, mammalian eggs with no yolk are much
smaller than those of non-mammals. The eggs of certain birds
may have a diameter of several centimeters. The sperm of a fruit
fly is remarkable, see Figure 1.1 for an illustration.

1.1.1   The Contents of a Cell

All cells contain a viscous liquid called the cytoplasm in which
are embedded a number of mini-organs called organelles
(see Figure 1.2). They make various contributions to the

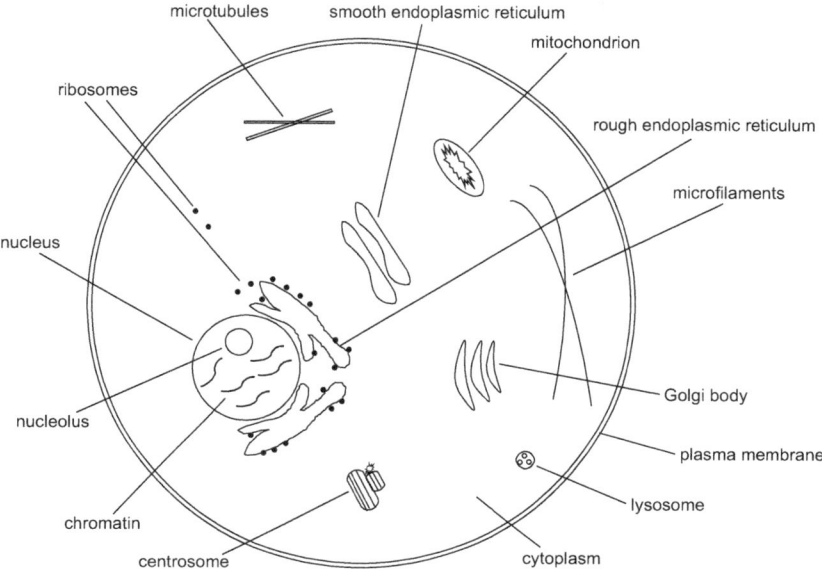

Figure 1.2　Cross section of an animal cell showing a number of organelles. A plural designation indicates there are more than one, sometimes many, present in the cell.

workings of the cell and to the well-being of the animal containing them. A plasma membrane encloses the cell.

Some organelles are particularly relevant (although all are important!) to genetics and a brief description of these follows:

- **Ribosome:** Millions in each cell, both free in the cytoplasm or attached to the rough endoplasmic reticulum. There are various types, but each consists of a complex of ribosomal ribonucleic acid (RNA) and proteins. They aid in the conversion of deoxyribonucleic acid (DNA) into proteins.
- **Endoplasmic reticulum:** There are two types. The smooth type is the site of lipid formation and poison detoxification. The rough type is embedded with ribosomes and is the site of protein synthesis. Both are near and sometimes merge with the nucleus.
- **Golgi body:** Often close to the endoplasmic reticulum, it modifies and sorts proteins for transport to various parts of the cell, particularly the cell membrane.

- **Lysosome:** This contains digesting enzymes that destroy bacteria and cell debris.
- **Centrosome:** In the cytoplasm region surrounding a pair of centrioles.
- **Centriole:** Consists of nine bundles of microtubules, which aid cell division.
- **Microtubule:** Consists of many subunits of tubulin. They maintain cell shape and move fluids over cell surfaces.
- **Microfilament:** Consists of polymerized actin. A few nanometers in diameter, many of them maintain cell structure and movement.
- **Nucleolus:** A small, dense round body within the nucleus but not membrane-bound in which ribosomal RNA transcription occurs.

The nucleus and mitochondrion are of particular interest to the geneticist and their salient features are shown in Table 1.1.

Table 1.1   Important properties of the cell nucleus and mitochondrion.

Nucleus	Mitochondrion
Size and containment	
The largest organelle, diameter around 10–30 μm and is about 5–10% of the cellular volume. Its shape often parallels that of the cell. It is observable with a microscope. The cell envelope is a multi-porous double-layered membrane.	Range of sizes, average is 1–10 μm in length and 0.5–1.0 μm in diameter. An outer membrane filters out large molecules and an inner membrane allows passage of only certain molecules.
Number in cell	
Usually only one or a few depending on cell type. In humans, there are none in red blood cells and many in bone and cancer cells.	Range from a single large one to a whole network of smaller oblong-shaped. A typical animal cell will contain around 1000 mitochondria.
Contents	
A semi-fluid liquid (nucleoplasm), chromatin and nucleolus. It contains over 98% of the cell's DNA.	Fluid matrix contains water, many enzymes and ribosomes, and a very small % of the cell's DNA.
Function	
Coordinates cell activities, including growth, division and protein synthesis.	Plays a central role in apoptosis and aging. It is the source of most of the adenosine triphosphate (ATP) needed for metabolism.

1.1.2 The Differentiation of Cells: Stem Cells

There is a series of complex changes in going from the single cell and finishing with the different specialized cell types. This is termed cell differentiation.

- The union of the sperm and the egg cell produce a starter cell termed a zygote (see Figure 1.3). In animals, this fertilized egg is almost perfectly round. It divides rapidly by mitosis and, in doubling, continuously forms 2, 4, 8, 16 and so on cells. The zygote becomes the embryo after the first zygotic division and the fetus after the eighth week.
- The fertilized egg and the first few cells are totipotent stem cells (see Figure 1.3). These are capable of forming any of the cells of the body, as well as the placenta and embryo. These cells start to become specialized, and in humans become a clump of 10–20 cells in the inner cell mass (ICM) within the globular blastocyst (around 150 cells) after about 3 to 5 days.

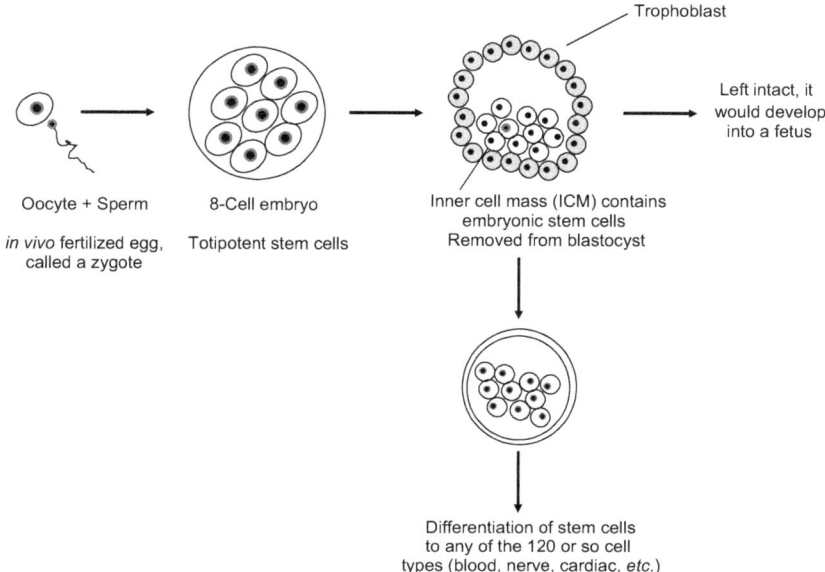

Figure 1.3 The generation of totipotent and pluripotent stem cells.

- The ICM yields the pluripotent or embryonic stem cells. These can also form any of the 200 cell types of the body but not the placenta nor the embryo. They can then propagate themselves for long periods. They create most of the cells and tissues that make up a body, such as muscle and nerve cells, skin and hair, *etc.* (see Figure 1.3).
- Non-embryonic tissue-specific stem cells in adults appear throughout the body in small amounts and are used to replace non-functional or dead cells. Outside of the US, stem cells have been used in regenerative medicine to help repair damaged and diseased cells in people who, for example, have suffered massive strokes. Adult stem cells can usually only generate the type of cell from their tissue of origin. Those from bone marrow can, therefore, only generate red or white blood cells. This fact has been used for years in bone marrow transplants to treat certain bone/blood cancers.

Normal somatic cells, *e.g.*, skin tissue or blood, can now be converted genetically into pluripotent stem cells, termed induced pluripotent stem (iPS) cells. Their use circumvents the ethical problem of human embryo destruction necessary in producing embryonic stem cells. iPS cells are likely to be important in studying a number of diseases, type 1 diabetes and osteoarthritis, for example. Stem cells feature in Chapter 8.

1.1.3 The Lifetimes of Cells

In humans, the average number of times the cell population doubles by division before the cells stop dividing (senescence) and finally commit suicide (apoptosis) is 40–50 times. This means something over 2^{40}, which suggests that 10^{12} cells will be formed in that time. There appears to be a rough correlation between the average lifetime of an animal and the doubling number (see Section 3.2.2).

The natural lifetime of the different cell types in an animal varies widely. In humans, skin epidermis cells are replaced every 10–30 days. The lifetime of a liver hepatocyte cell is some 0.5–1.0 year, while the central nervous system cells last a lifetime.

1.2 DNA AND RNA

1.2.1 DNA

Deoxyribonucleic acid or DNA (surely one of the most famous abbreviations) comprises two extremely long and similar strands, which are weakly bound together. Each strand consists of a linear array of poly-sugar–phosphate molecules, with one of four bases attached to every sugar molecule. This one base can weakly bind to a particular base protruding from the other strand. A fragment of the DNA structure is shown in Figure 1.4.

Adjacent deoxyribose sugars are attached to each other using two of the oxygens of a PO_4^{3-} ion, one oxygen directly binding to the 3′ position of one sugar and the other oxygen binding to the 5′ position of an adjacent sugar (Figure 1.4).

1.2.2 The Interaction of the Bases between the Two Strands

Four different bases are involved in DNA, as shown in Figure 1.5.

The bases are strongly (covalently) bound by one of their ring nitrogens to the 1′ position of the sugar. The association of the base on one strand is by hydrogen bonding to the base on the other strand. This is not a haphazard arrangement, for a two-ringed adenine base (A) is always weakly attached to a one-ringed thiamine (T) by two hydrogen bonds. A two-ringed guanine base (G) is always associated with a one-ringed cytosine (C) by three hydrogen bonds, and is therefore the slightly stronger association (Figure 1.5). Seven of these interactions are shown in Figure 1.4. The association of one base on one strand with the corresponding base on the other strand is necessarily weak. This means that the strands of the DNA can be relatively easily and reversibly separated, by heating, for example (termed melting). This is an important property for many reasons. The whole arrangement can be viewed as a flexible ladder-like structure. The two extremely long strands of the poly-sugar–phosphate entity are the upright lengths of that ladder and the complementary bases are the cross rungs. The ladder does not, however, remain flat but is twisted multiple times to form a long spiral with the two legs (strands) aligned in opposite directions (antiparallel). This right-handed double helix is the usual (but not only) biological conformation. The helix undergoes one complete

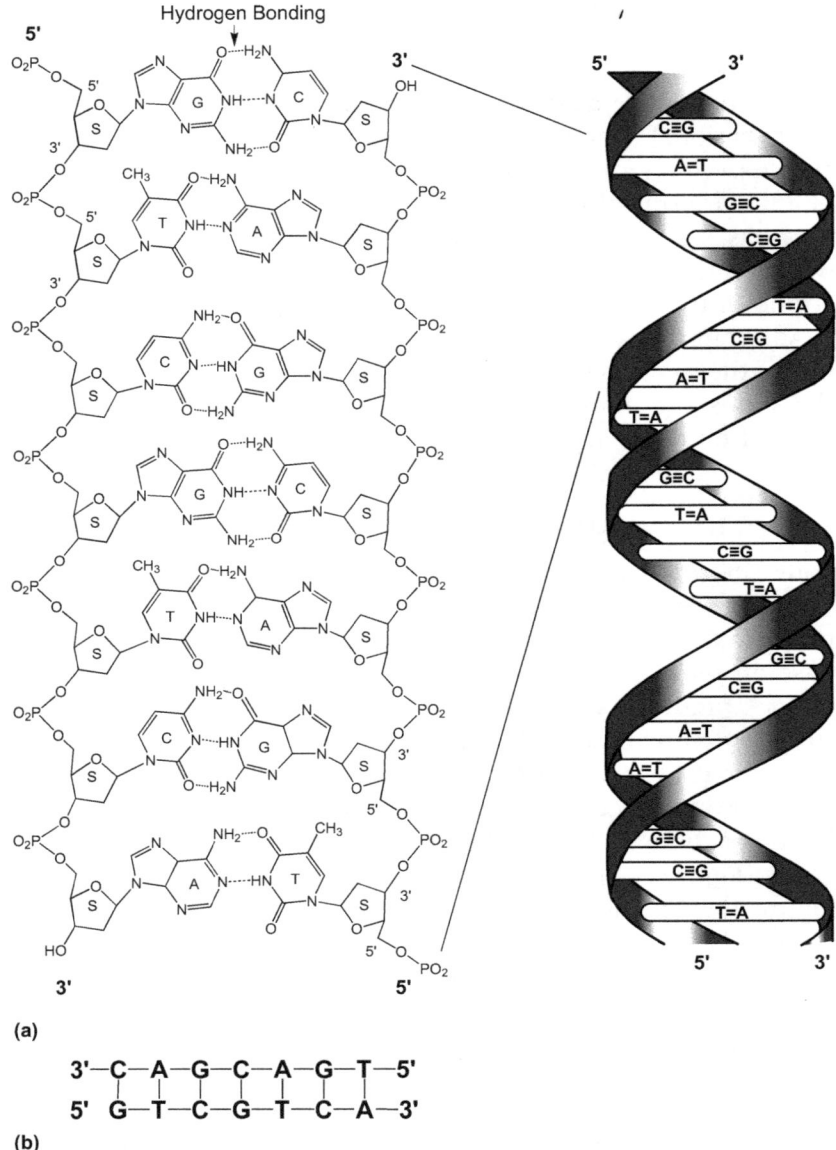

(a)

(b)

$$3'-C-A-G-C-A-G-T-5'$$
$$5'-G-T-C-G-T-C-A-3'$$

Figure 1.4 The structure of a short segment of DNA shown in (a) spiral and (b) linear arrangements. The four bases (adenine, cytosine, guanine and thymine) and sugars represented by A, C, G, T and S, respectively, have the structures shown in Figure 1.5.

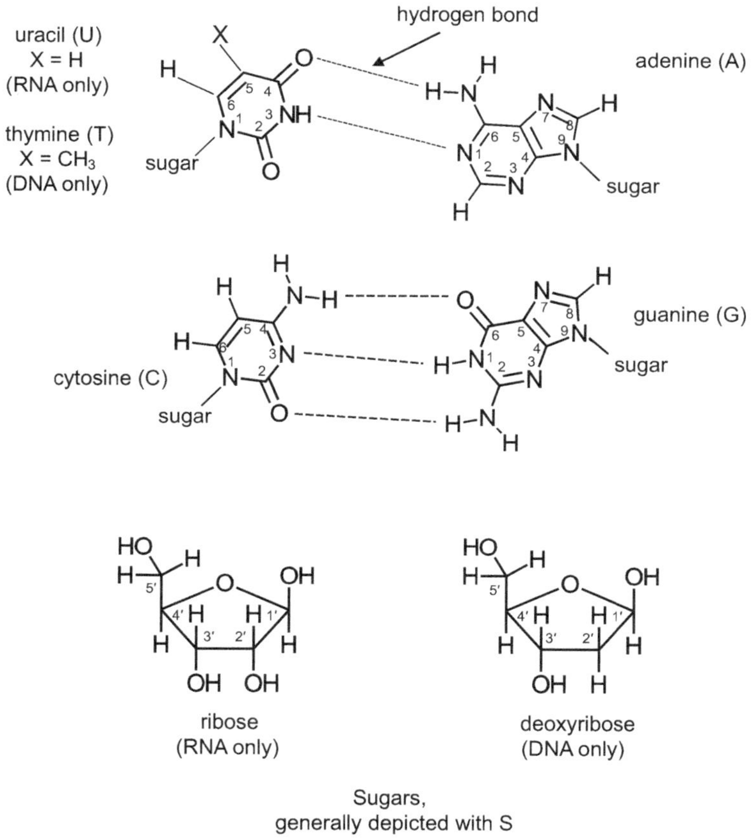

Figure 1.5 The structures of adenine (A), cytosine (C), guanine (G), thymine (T) and uracil (U) bases, and ribose and 2-deoxyribose sugars (S) in RNA and DNA. The specific hydrogen-bond associations of the bases in DNA are shown.

revolution with each 10 base pairs (bp). Each base pair is therefore rotated 36° with respect to its neighbors. Bearing in mind that DNA is a spiral arrangement, it is usually represented as a straight, flat segment, as shown in Figure 1.4(b).

1.2.3 Nucleotides

Nucleotides are the monomeric units of nucleic acids, *e.g.*, DNA and RNA, and consist of a nitrogenous base, a five-carbon sugar and one or more phosphate groups. The molecules shown in

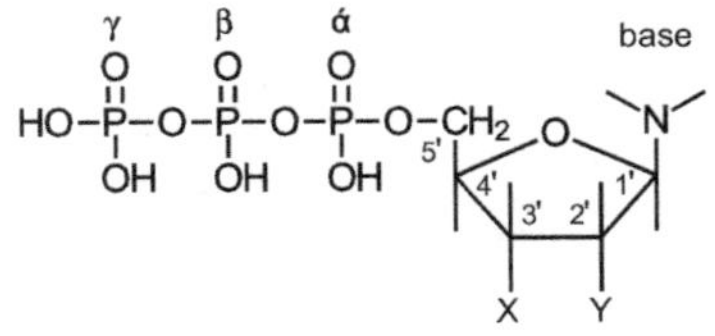

Figure 1.6 structures:

Structure A: If X = HO and Y = H, the structure is dNTP

Structure B: If X = H and Y = H, the structure is ddNTP

Structure C: If X = HO and Y = HO, the structure is NTP

Figure 1.6   The three nucleotides NTP, dNTP and ddNTP. These differ in the 2′ and 3′ substitution in the ribose sugar. The bases are shown in Figure 1.5.

Figure 1.6 play an important role in the manipulation of DNA and RNA, both within the cell and externally by the scientist.

Deoxynucleotide triphosphate (dNTP; Figure 1.6A) is sometimes written as deoxyribonucleoside triphosphate and even simply as a nucleotide when the context is understood. The letter d signifies deoxy, with an –OH group absent from the 2′ position of the sugar. The N signifies any of the four bases. The nucleotide dNTP features in DNA replication (see Section 1.2.4) and the polymerase chain reaction (PCR; see Section 2.3 in Chapter 2). In the four dideoxynucleotide triphosphates (ddNTP; Figure 1.6B), there are now no –OH groups on the sugar. Sanger sequencing (see Section 2.1.1 in Chapter 2) employs ddNTP. Four molecules, collectively called nucleotide triphosphate (NTP; Figure 1.6C), belong to the RNA group, where –OH groups at both the 2′ and 3′ positions of the sugar and uracil replace the thymine used in DNA. NTP is employed in transcription (see Section 3.1.2 in Chapter 3) and in, for example, CRISPR/Cas9 gene editing (where CRISPR is clustered regularly interspaced short palindromic repeat and Cas9 is CRISPR-associated protein; see Section 8.4.9 in Chapter 8).

1.2.4   The Elongation of the Strand

The process of strand elongation is shown in Figure 1.7.

The successive attachments must proceed in the 5′ to 3′ direction since a free 3′ –OH group on the sugar is required, as

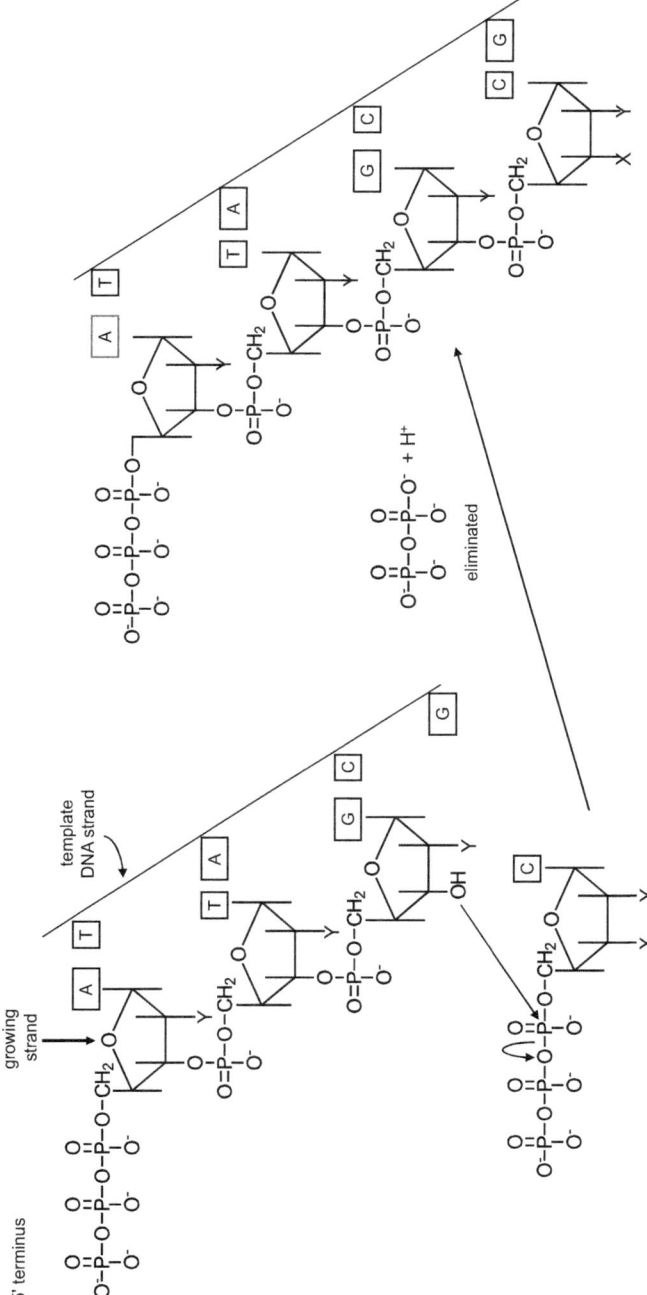

Figure 1.7 The elongation of the DNA strand using deoxycytidine triphosphate, dCTP (X = OH, Y = H), to attach cytosine (C) to guanine (G) on the template DNA strand. Adjacent sugars are attached to each other using the nucleotide with release of pyrophosphate, $P_2O_7^{4-}$, and H^+ ions, placing the base in a position enabling a hydrogen bond to the complementary base on the other strand.

shown in Figure 1.7. Thus, on the left side of Figure 1.4, the successive attachments would have proceeded downwards, and in the complementary strand (Figure 1.4(a), right) the successive attachments would proceed upwards in the $5'$ to $3'$ direction. This arrangement might appear trivial but is in fact extremely important. It is invoked when single-stranded DNA is duplicated (DNA replication), when DNA is sequenced (DNA sequencing), when DNA damage is repaired and when RNA is attached to long (transcription) and short (translation) single-stranded DNA.

1.2.5   Cellular DNA

There are two sources of DNA within the cell, namely in the nucleus (nuclear DNA, nDNA) and in the mitochondrion (mitochondrial DNA, mtDNA). Nuclear DNA provides the vast majority of cellular DNA in mammals, being responsible for most of the cell functions. The number of base pairs (*i.e.*, the A–T or C–G interactions) in nDNA varies widely amongst animals. The amounts may be as small as 137 million in fruit flies to as many as several billion bases in mice and humans. The packing of this large bulk into the nucleus of the cell presents a problem, which we consider in Section 1.3. Normally, all (unmodified) DNA contains the same bases, the same arrangement of base pairing and the same double helix in all types of cells and, even more significantly, in all animals. This is a remarkably constant factor. Imagine trying to understand animal genetics were this not so!

DNA appears in a number of guises and functions, which are explored throughout the book. One particularly interesting and important one involves reversible methylation of a cytosine base at position 5 (Figure 1.5), forming 5-methylcytosine. This does not jeopardize the fundamental structure of DNA but can have far-reaching effects on the expression of the gene (see Section 7.4 in Chapter 7). It is estimated that, in mammals, 60–90% of all cytosine that is immediately adjacent to a guanine base is methylated.

1.2.6   Blocking DNA Replication in Pathogen Contaminants of Blood

The basic elements of blood are white blood cells (leukocytes), red blood cells (erythrocytes) and platelets (thrombocytes), which are all confined in membranes and contained in a clear extracellular fluid (plasma) consisting mainly of water. Only the white blood cells contain a nucleus and DNA. The absence of DNA in the other blood components has been cleverly exploited to inactivate a broad spectrum of pathogens (bacteria, viruses and protozoa), which might, in very small amounts, infect a blood sample—a small molecule, amatosalen, is added to blood platelets or plasma. It penetrates cells and pathogens and specifically intercalates (a weak interaction) into the helical regions of any DNA or RNA. Upon ultraviolet (UV) illumination, much stronger intrastrand crosslinking occurs, thereby blocking DNA replication or repair in the pathogen, meaning it can no longer multiply. The platelets or plasma are unaffected. Thus, the process *intercepts* any infection in the platelets or plasma before reaching a patient awaiting a transfusion. The process is not applicable to red blood cells since these are opaque to the UV frequencies used. A recent modification of the method uses a chemical to crosslink and inactivate DNA without UV illumination so that red blood cells can also be treated. The testing of platelets for some bacteria may take a critical two or so days when their shelf life may only be five days. The intercept process takes only minutes for plasma and hours for platelets. Chikungunya fever is caused by a virus for which a blood screening test is not available. It is treatable by the intercept system. This proved invaluable during a recent outbreak in Puerto Rico. The system has been used in transfusions in Europe for a few years.

1.2.7 RNA

In many respects, the structure of RNA must resemble that of DNA since it undergoes the same type of interactions that are seen with DNA. It can even associate with single-stranded DNA by using the hydrogen-bonded base–base interactions observed in DNA. RNA does, however, differ from DNA in important respects, as follows.

1. The binding of the base and the phosphate entity with the sugar in RNA are at the same positions as in DNA. RNA is normally single-stranded. It can, however, form short double-stranded structures by binding to itself (Figure 1.8).

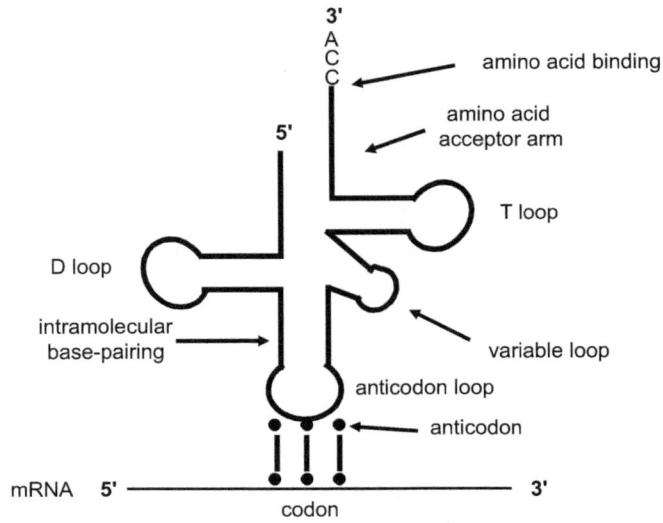

Figure 1.8   A simplified representation of transfer RNA (tRNA), which typically contains 75–95 nucleotides. These are usually folded into a characteristic cloverleaf structure. Single-stranded RNA is present at the 3′ end and in the four loops. Double-stranded, self-binding RNA is used in the four stem regions. A number of bases are modified from their normal structures. tRNA mediates the synthesis of polypeptides using messenger RNA (mRNA; see Section 3.1.5 in Chapter 3).

2. The sugar in RNA is ribose with an –OH group replacing the H atom at position 2 in the sugar in DNA (see Figure 1.5).
3. Thymine (T) in DNA is replaced by uracil (U) in RNA. This entails replacing a methyl group in thymine by a hydrogen atom in uracil at position 5 (see Figure 1.5). The interactions now are adenine with uracil (in RNA) and cytosine with guanosine.

RNA, like DNA, appears in a number of forms and functions, and particularly in gene expression (see Section 3.1 in Chapter 3).

1.2.8   Oligonucleotides

Oligonucleotides are short DNA or RNA molecules that are often 15–25 nucleotides in length. In nature, they are found both single and double stranded, and function in the regulation of gene expression (see Section 3.1.8 in Chapter 3). Synthetic oligonucleotides of varying lengths find wide applications in the laboratory. They are designed to hybridize (pairing of complementary strands) to DNA or RNA sequences, for instance as primers in PCR (see Section 2.3.2 in Chapter 2), in DNA sequencing (see Section 2.1.1 in Chapter 2), gene cloning screening (see Section 2.2.1 in Chapter 2), in forensics (see Section 8.1.1 in Chapter 8), as molecular probes (*e.g.*, in FISH; see Section 1.4.3) and other areas. Chemical modification of the ribonucleic structure shown in Figure 1.4, while still retaining the bases, produces antisense oligonucleotides. These can be used to match a region in sense messenger RNA (mRNA), where a damaging mutation resides and negate it. This is the basis of their potential use as drugs to treat diseases, termed antisense technology (for more details, see Section 8.5 in Chapter 8).

1.3 DNA PACKAGING IN THE CELL

1.3.1 Nuclear DNA

There is a considerable problem in fitting the nuclear DNA into the nucleus of a cell. Chromosome 1 is the longest human chromosome and contains about 250 million bp. If it is assumed that the average length of a base pair in the DNA helix is 0.34 nm, then this amounts to an extended length of $0.25 \times 10^9 \times 0.34 \times 10^{-9}$, *i.e.*, approximately 0.085 m of DNA. This has to be squeezed into a nucleus with a diameter of 3–10 μm, thus representing an obvious problem. The insertion of DNA into chromosome 1 is accomplished using a series of compactions until the desired reduction in size is achieved (Figure 1.9). Since chromosome 1 is 10 μm in length, this amounts to a packing ratio of $8.5 \times 10^4/10 = 8.5 \times 10^3$, where the packing ratio is the length of DNA divided by the length into which it is packaged. Since the total number of base pairs in the human, which consists of 23 chromosome pairs, is around 6 billion, the challenge of packing the DNA is formidable.

1.3.2 Compaction Details

There are distinct hierarchies of organization of DNA during its compaction to chromosome (Figure 1.9).

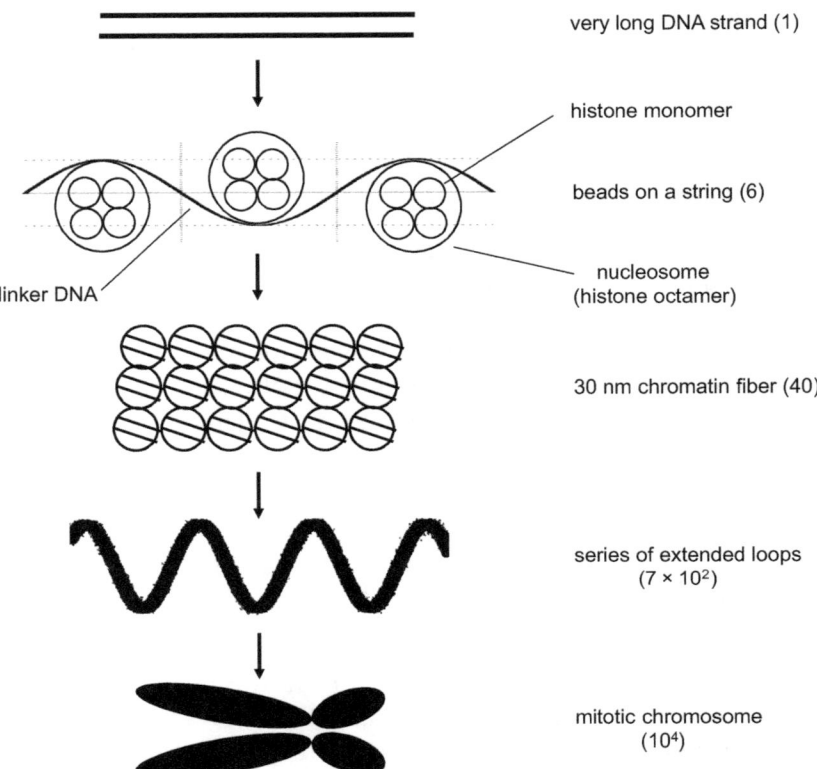

very long DNA strand (1)

histone monomer

beads on a string (6)

nucleosome
(histone octamer)

linker DNA

30 nm chromatin fiber (40)

series of extended loops
(7×10^2)

mitotic chromosome
(10^4)

Figure 1.9   Important structures and their packing ratios (in parentheses) used in transforming the DNA strand to the chromosome, the final compact form. These structures are strung together during compaction.

The first level of packing is an important one. A sequence of 146 bp in the DNA strand is wrapped almost twice around an octamer of histone proteins, much like a thread of cotton around a spool. Histones are small, positively charged proteins that have an affinity for the negatively charged DNA. As well as playing a structural role in compaction, histones may be chemically modified and play an important part in the expression of the chromosome gene (see Section 7.4 in Chapter 7). The histone octamer is termed a nucleosome and is joined to the next nucleosome by linker DNA, ranging from 8–114 bp, depending on the animal and even the cell type within one animal. This gives a "bead on a string" appearance. The nucleosome is a fundamental unit of chromatin, which in turn is a functional subunit of a chromosome.

A string of tightly packed nucleosomes coil in a helical structure called a solenoid or 30 nm fiber. At this stage the packing ratio is still small (around 40).

The most marked packaging subsequently occurs after the fiber is organized into looped domains, which are held together by a matrix of scaffolding proteins. This occurs in the interphase of the cell cycle (see Section 3.2 in Chapter 3) when the cell is not dividing.

About 18 of these loops further condense to form minibands (not shown in Figure 1.9). A large number of minibands constitutes the tangled mass of chromatin, which is a complex of DNA plus proteins (mostly histones).

Chromatin is duplicated during the interphase of the cell cycle and then further condensed during mitotic metaphase (see Section 3.4 in Chapter 3) to give the classical chromosome shape, visible by light microscopy. The overall packing ratio from linear DNA into interphase chromatids (see Section 3.4 in Chapter 3) is about 10^3, and to the final mitotic chromosome is somewhat more than 10^4. The other 45 chromosomes that make up the human DNA are formed in a similar way, leading to an overall packing ratio of around 4×10^5. This allows all the chromosomes to fit into the nucleus.

Chromatin may be present as either euchromatin or heterochromatin, both consisting of very many nucleosomes. Euchromatin is an open form of chromatin and loosely wrapped around histones making it more accessible to machinery that allows, for example, the DNA to be transformed into proteins. Heterochromatin contains, on the other hand, tightly coiled DNA and this means that only parts are genetically active.

1.3.3 Mitochondrial DNA

The number of base pairs (16 500 bp) in a single copy of mitochondrial DNA is in the form of a small ring and is remarkably constant in all animals. This represents only about a 5.6 μm length of DNA, which must be inserted into a few micrometers of mitochondrion. A little problem therefore arises in the packaging of this small amount of DNA into the mitochondrion of the cell. Even now it appears that a protein assists in this operation, similar to that played by histones in nuclear DNA.

1.4 THE CHROMOSOME

During the interphase of the cell cycle the chromatid resulting from the compacting of the DNA strand is duplicated (see Section 3.3 in Chapter 3). The two identical (sister) chromatids are termed a chromosome. The chromatids are held together at a constriction point called the centromere (Figure 1.10), which separates the shorter p-arm from the q-arm, and is the point of spindle attachment during mitosis and meiosis. The centromere may be around the middle (metacentric), somewhat off center (submetacentric) or near the end (acrocentric) of the chromosome. The acrocentric chromosomes have a nearly terminal centromere and the shorter p-arm is hard to observe.

In most animals, the female and the male each contribute a chromosome to give a pair of chromosomes and a total of four chromatids (Figure 1.10). The maternal and paternal chromosomes are not identical (see Section 1.6.2) but resemble each other closely in important characteristics. Thus, they have the same length, form the same light and dark patterns on staining (see Section 1.4.3), and have the same genes for the same

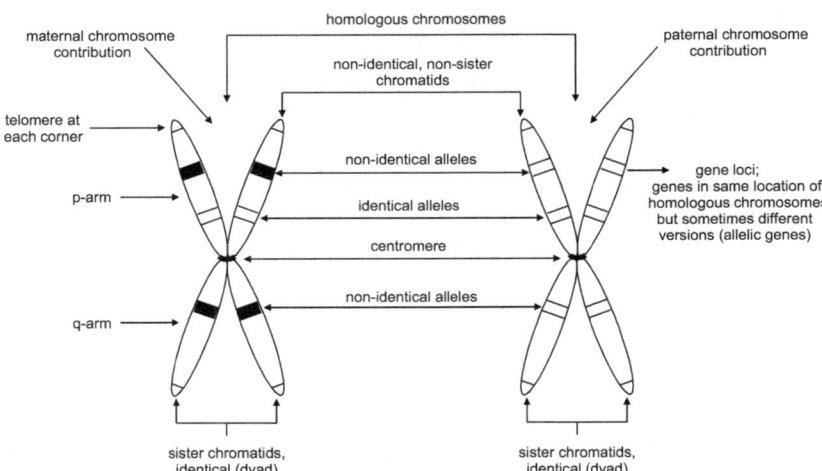

Figure 1.10 Various features of a pair of chromosomes.

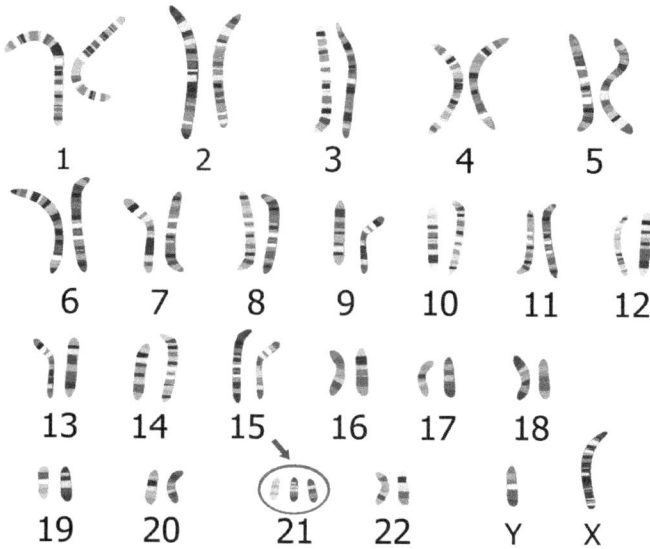

Figure 1.11   Karyotype of the chromosomes of a human male with trisomy 21.
© Zuzanae/Shutterstock

inherited characteristics and in the same order. The exception to this generalization is the two male chromosomes, which differ markedly in size (Figure 1.11).

Also shown in Figure 1.10 are the telomere regions, which lie at the tips of a chromosome and whose function, in part, is preventing the "fraying" of the chromosome, much like the plastic tips of a shoelace. It is worth noting that a chromosome is often depicted as a single rod-like structure, thus disguising the presence of two sister chromatids but still showing the centromere connecting point.

1.4.1   Chromosome Numbers

All animals have a characteristic number of chromosomes in each cell termed the diploid $(2n)$ number. The diploid number in different animals varies a good deal, sometimes within the species. There are, for example, a variety of chromosome numbers $(2n)$ associated with monkeys, ranging from 34 (in the spider monkey) through 42 (in the rhesus monkey) to as many as 72

Table 1.2 Number of chromosomes (2*n*) in the cell nucleus for animals featured in the book.

Animal	(2*n*)	Animal	(2*n*)
Alligator	32	Hare	48
Bear (polar, black, *etc.*)	74	Hamster (common)	22
Bear (giant panda)	42	Horse (domesticated)	64
Bee (honey), female	32	Horse (Mongolian, wild)	66
Bee (honey), male	16	Human	46
Butterfly (monarch)	29–30	Kangaroo	16
Butterfly (bombyx)	28	Monkey	34–72
Cat	38	Mouse	40
Cattle	60	Mule	63
Chicken	78	Pig	38
Dog	78	Platypus (duck-billed)	52
Donkey	62	Rabbit	44
Elephant	56	Rat	42
Fly (fruit)	8	Rat (naked mole)	50–60
Fly (house)	12	Sea urchin	36
Goat	60	Sheep	54
Great apes	48	Woolly mammoth	58
Guinea pig	64		

(in the old world monkey). Many animals have numbers around that of humans (*i.e.*, 46), see Table 1.2.

1.4.2 Chromosome Types

Most animals have a set of chromosomes, which generally vary relatively little in size. Between animals there is, however, a wide variation in their characteristics. Microchromosomes are very small (dot-like) and often indistinguishable, and are found, for example, in chickens and some reptiles, as well as fish, but do not exist in mammals. All the mice chromosomes are acrocentric with a very little p-arm. The vast majority of the chromosomes (numbered 1–22 in humans, for example) are non-sex and called autosomes. Each diploid organism will have one set of autosomes from the male and a set contributed from the female parent. A sole remaining pair are the sex chromosomes. These are involved with the sexual organs and therefore the reproductive process. With most animals, the sex of an offspring is determined by the sex chromosomes. With placental mammals and marsupials (as well as several species of turtles and lizards), the female has two sex chromosomes, designated XX, and the male has two dissimilar chromosomes, designated XY. The male and female each contribute one sex chromosome to their off-spring (see Section 5.2 in Chapter 5).

1.4.3   Chromosome Imaging

How do we arrive at the numbers, sizes and shapes of chromosomes in a cell? A variety of tissue types can be used to examine this since the cells from any tissue normally contain an identical composition of DNA and chromosomes. The cell is first stopped from dividing by locking it *in vitro* at the mitotic metaphase. This is achieved by adding colchicine, an alkaloid that by binding to the tubulin dimer in the mitotic spindle (see Section 3.4 in Chapter 3) inhibits its formation. Chromosomal proteins are enzymatically degraded using trypsin and the specimen is then spread on a slide, stained with a dye (Giemsa is a favorite), examined under a light microscope (1000-fold magnification) and photographed. Dye binds to the DNA, giving so-called G-bands. In every chromosome in the animal, the dye adheres particularly strongly to heterochromatin, giving dark bands, and much more weakly to euchromatin, giving lighter bands. The result from a human tissue is shown in Figure 1.11.

The "wriggly-worm" appearance in the karyotype arises from the flattening of the chromosomes on the slide. It can be converted to a neater, graphical representation, termed the karyogram or ideogram. This shows the chromosomes in humans arranged in order of decreasing size and numbered 1 to 22, plus the pair of sex chromosomes, in this case X and Y in a male. There are some hundreds of bands in the complete set of 23 human chromosomes in the metaphase of mitosis, each containing a number of genes that can be distinguished and assessed by experts (see Figure 1.12).

FISH and SKY

Newer techniques use a series of fluorescent dyes that bind to specific regions of the chromosome or to specific chromosomes. In fluorescence *in situ* hybridization (FISH), a target sequence of DNA (in a tissue or a chromosome) that has been denatured to a single strand is mixed on a glass slide with a DNA or RNA

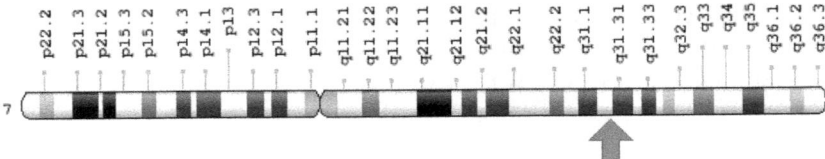

Figure 1.12   The CFTR gene, a mutation of which may cause cystic fibrosis, is located in humans on chromosome 7 on the long (q) arm in region 3, band 1 and sub band 2 (designated by the arrow), and thus assigned the position 7q 31.2. The gene designation in the book will, in general, be restricted to the chromosome number and either the short (p) or long (q) arm, *i.e.*, in this case 7q.

oligonucleotide probe that matches the target sequence by hybridization, *i.e.*, the bases are complementary. If the probe is labeled with a fluorochrome (a fluorescent dye), the site of the binding can be detected with a fluorescent microscope. A normal chromosome will have a uniform color, whereas any disturbed chromosome will have a striped appearance. The whole process has been likened to finding a needle (the target DNA) in a haystack (the chromosome) using a powerful magnet (the probe). In a similar way, in spectral karyotyping (SKY) all 23 chromosomes are treated with probes that bind specifically to one of the chromosomes by hybridization. These fluorescent probes produce different colors by Xenon-light activation so that each chromosome shows a different color, which is distinguishable by eye or with the help of a computer. In this way, a large chromosome aberration, for example, a translocation (see Section 4.4.3 in Chapter 4), shows as a "mixed" color on two chromosomes, which is more easily discerned than by G-banding karyotyping analysis, although both methods are often used in conjunction.

1.4.4   The Value of Karyotypes

We consider three examples of the type of information that may be gleaned from an examination of the karyotype of a human or animal.

1.4.5   Giant Panda and Other Bears

Is the giant panda a member of the bear family? When one considers that most bears (polar, black, brown, *etc.*) have 74 chromosomes in a cell, whereas the giant panda has only 42, one might have some reservations. However, examination of the karyotypes of the two animals shows that the smaller number with the giant panda results from it having a number of larger chromosomes found as a combination of two chromosomes in the bear species. For example, chromosomes 1, 2 and 3 in the giant panda have almost identical banding characteristic to the sum of chromosomes 2 and 3, 1 and 9, and 6 and 16, respectively, in the brown bear. One can therefore conclude that both belong to the same family and that fusion of chromosomes took place during the panda lineage.

1.4.6 The Chimpanzee and Human

A similar type of situation appears in an examination of the ideograms of chimpanzees and humans. Chimpanzees, gorillas and orangutans have 24 pairs of chromosomes. Humans have one pair less (although, until 1955, they were also believed to have 24 pairs). Two separate chimpanzee chromosomes, namely 12 (2p) and 13 (2q), when laid end-to-end, create an identical staining pattern to human chromosome 2. The other chromosomal-banding patterns of the two animals are very close. These facts strongly suggest a common ancestry for apes and humans.

1.4.7 Human Chromosome 21

The karyotype can be an invaluable diagnostic tool for animal traits and disorders. The presence of a full or partial extra copy of chromosome 21 in humans, showing three rather than the usual two chromosomes, is easily seen in the karyotype (see Figure 1.11). This presages trisomy 21 (Down syndrome). Such a disorder can be shown even in human fetal DNA using non-invasive prenatal testing. Most cell-free DNA in plasma is maternal, but cell-free fetal DNA is present to the extent of 4–15% and detectable after about 10 weeks. A pregnant woman carrying a fetus with trisomy 21 will have a slightly higher amount of chromosome 21 cell-free fetal DNA than expected in the plasma. The very small difference can be resolved, for example, by next generation sequencing (see Section 2.1.4 in Chapter 2). Other aneuploidies (see Sections 5.2.3 and 5.2.4 in Chapter 5) can be diagnosed in this way, including trisomy 13 and 18, 45X and others, as well as the sex of the fetus without resorting to invasive prenatal diagnosis.

1.5   DNA SEQUENCES

A DNA sequence is the precise order of nucleotide residues within a DNA strand. The arrangement of DNA in a chromosome is not completely haphazard. Chromosomes of animals contain very many repeated DNA sequences. Their contents may vary and the length of the repeat sequence may range from one nucleotide to a whole gene.

The genome (all the DNA; see Section 1.7) of the annual fish *Nothobranchius furzeri* ($2n = 38$) has the highest percentage (45%) of repeat segments of any fish so far examined. This may be related to its short life (3–4 months in captivity), which is the shortest of any vertebrate.

There are basically two dispositions of the DNA repeats. They may be tandem repetitive or interspersed repetitive.

1.5.1   Tandem Repetitive DNA

These repeats are usually identical and adjacent to one another. They comprise 10–15% of mammalian DNA. Tandem repetitive DNA is often located near or in the centromere or telomere regions of the chromosome, which it may structurally protect.

Satellite DNA has some 10^6 copies of short nucleotide sequences. It is termed satellite DNA because it shows a different band near the main band during ultracentrifugation. The satellite segments have a different base composition and different density than the remainder of the DNA. If the satellite DNA is $(A + T)$ rich, it shows up as lighter density bands compared with the band associated with the majority of the DNA, while if it is $(G + C)$ rich, it shows up as higher density bands.

Another example of tandem repetitive DNA involves variable number of tandem repeats (VNTR), also termed minisatellites. The length of the repeat unit may be 10–100 bases but can span a length up to 20 kb. VNTR occur much less often than a final example of repeats, namely short tandem repeats (STR), termed microsatellites. These are short sequences of 1–13 (often 2–6) bases and usually less than 150 bases overall in length. STR are found in all animals and appear randomly amongst all the chromosomes, usually outside the DNA coding (gene) area and are therefore located in the vast majority of the genome. The most abundant type of STR is a trinucleotide repeat. Normally, these triplets are repeated for anything from 5 to 50 times. When this number of triplets is exceeded, perhaps reaching 40–3000 repeats, either a mutant (toxic) protein forms or there is an altered gene expression. Both events may result in disorders, the severity of which increases as the expansion increases (see Section 4.4.1 in Chapter 4). The human genome contains many thousands of STR in all the chromosomes. This type of sequence is the same and appears in similar chromosome positions for all people, but (and very importantly) the number of copies might vary from as few as 7 to more than 40 from one person to another and from one chromosome to another chromosome (polymorphism). This enables DNA fingerprinting (see Section 8.1 in Chapter 8) to be so effective.

1.5.2 Interspersed Repetitive DNA

These are repeat units scattered randomly through the genome and they are very similar but not identical (cf. tandem repetitive). They are confined to mammals and comprise 25–40% of their DNA. They include long interspersed nuclear elements (LINES), which comprise more than 5000 base segments in more than 10^4 copies, and short interspersed nuclear elements (SINES), which have fewer than 500 base segments and 10^5–10^6 copies.

The SINES group includes the Alu family, which comprises about 10% of the human genome and is its largest DNA repeat component. It is exclusive to primates. The Alu element is about 300 bp, with a highly repetitive sequence rich in CpG islands. (The p symbol means that the C and G nucleotides are joined by the usual phosphodiester bond.) Since a large fraction of Alu elements are in the intron region (see Section 3.1.3 in Chapter 3), it is not involved in making protein. Unchecked, Alu elements can be irreversibly inserted into a coding region of the genome, affecting transcription and disruption of gene function and eventual cell death. This can be held in check by methylation of the CpG islands (epigenetics; see Section 7.4 in Chapter 7). Some severe diseases, such as breast cancer and hemophilia, have been associated with Alu element insertion into the coding region of a gene. Alu elements in primates are useful in the study of their ancestry since insertion is very specific and primate species will only share that specific Alu insertion if they have a common ancestor. Interspersed repetitive DNA, as well as tandem repetitive DNA, feature heavily in epigenetics, in which methylation of LINES and Alu elements plays an important role in histone modification (see Section 7.4 in Chapter 7).

1.6   THE GENE

The gene is a sequence of nucleotides in a very small fragment of the DNA of a chromosome. It represents a small packet of information. The order of nucleotides in a gene provides a code for the synthesis of a cellular molecular product (see Section 3.1 in Chapter 3). A gene is responsible for implementing the very many physiological properties of an animal, *e.g.*, a particular characteristic (phenotype) of an animal, such as eye color in humans, wing structure in fruit flies, coat color in animals and so on. The gene occupies a specific position (locus) in the chromosome. This location is usually designated in relation to a particular light or dark band on a stained chromosome (Figure 1.12).

1.6.1   The Two Basic Classes of Genes

1. One class of gene, a structural gene, encodes proteins or RNA (*e.g.*, mRNA) not involved in gene regulation. They represent over 90% of the total genes in the human. They are expressed in transcription and translation processes (see Sections 3.1.2 and 3.1.5 in Chapter 3). It is estimated that anywhere between 250 000 to 1 000 000 proteins are encoded by the 19 000 genes in the human genome.
2. The second class, a regulatory gene, encodes regulatory proteins and, although only 2–5% of the total number of genes, is vital in controlling gene expression, which may control the expression of one or more structural genes. They encode proteins and a number of RNA molecules involved in gene expression regulation (see Section 3.1.7 in Chapter 3).

There are many genes contained in a dark or white band of a karyotype and many such bands in a chromosome. Many human genes share a very similar construction and function with genes in other animals. There is some variation in the number of genes in animals. The fruit fly has about 14 000 genes, but the water fly, which is barely visible, possesses 31 000! Many animals have a number clustered around the human figure. There is a relatively large range in the size of genes. In humans, the gene that encodes the beta chain of hemoglobin is about 16 000 bp, whereas the largest gene, responsible for the production of dystrophin, spans a mammoth 2.3 million bases. The genes still represent, however, a very small percentage of the DNA in a cell. The remainder, formerly unkindly termed "junk", is known to play an important role in controlling gene function.

1.6.2   The Gene and the Allele

In a diploid cell, *i.e.*, one with two sets of chromosomes, which is present in most animals, one chromosome of the pair originates from the mother (the maternal chromosome) and one from the father (the paternal chromosome). We know that these two chromosomes are homologous and can line up exactly so that corresponding genes match. The alleles are identical in the sister chromatids of each chromosome but not necessarily to the sister chromatids of the other chromosome of the pair (Figure 1.10). It is not easy to see how the exact line up could occur in the male sex chromosomes with a maternal-derived X and a paternal-derived Y that are so different in size (Figure 1.11). In fact, the X and Y chromosomes share a small region of homology, which have common genes and allow essential synapsis (matching) to occur between the two chromosomes during the metaphase of meiosis I (see Section 5.1 in Chapter 5). These are termed *pseudo* autosomal regions (PAR) and occur at either ends of the X and Y chromosomes.

The corresponding gene on homologous chromosomes will generate the same kind of phenotype but possibly in different ways, for example, eye color in humans, resulting in either brown or blue eyes; the wing structure in fruit flies, resulting in either long or short wings; the coat color in animals, resulting in either brown or black; a physical condition, either normal or a disorder; and so on. When the alleles (genes) differ, it may be by only one or a few DNA nucleotides.

1.6.3   The Gene and the Eye

Although it is known to be a simplification, let us assume that one gene (or allele) is responsible for eye color in humans, specifically brown or blue. Each parent contributes one chromosome and therefore one allele to the child. If both parents carry the allele for brown eyes, the child will have brown eyes. Similarly, if both parents carry the blue allele, the child will have blue eyes. In these instances the two genes are said to be homozygous. However, if one parent contributes a blue allele and the other parent a brown allele (heterozygote alleles), what will the child's eye color be? In many instances one allele is dominant over the other, recessive allele (see Section 6.2.1 in Chapter 6). In this case, the allele for brown eyes is dominant over that for blue eyes and will override it so that the child will likely have brown eyes. The term "likely" is used because of the simple assumptions. How else would you explain two parents with blue eyes, who should have a child with blue eyes, having a brown-eyed child? In fact, two genes close together on human chromosome 15 play major roles and a number of other genes have minor roles in deciding eye color (see Section 6.2.11 in Chapter 6).

A single gene appears to control the growth and development of eyes throughout the animal kingdom, despite the large differences in their eye construction. In humans, this gene, *PAX6*, is located on chromosome 11p. During embryonic development, the gene encodes a protein, PAX6, which is involved in many aspects of eye development. The protein attaches to specific regions of DNA and, in so doing, regulates the activity of other genes. Even a small change in the gene [over 250 different faulty versions (mutations) have been reported] can lead to a shortened and non-functional protein. This disrupts the formation of a normal eye and, instead, the iris or even the whole eye may be missing (aniridia). Only one of the two alleles of the gene need be mutated for its dominant presence to be felt. This is in contrast to the case with eye color (above), where the normal eye color (brown) is dominant and both mutant alleles (blue) must be present for a blue-eye inheritance. Homologs of the *PAX6* gene, *i.e.*, genes with the same structural features, have been found in rats, mice, zebra fish and fruit flies. The encoded proteins are all very much like PAX6, with as much as 90% similarity in their amino acid sequences. For the fruit fly, the defect in the corresponding gene, namely eyeless or the *Ey* gene, must be present in both alleles of the gene for the eyeless fly to develop. *The name of the gene is often based on the aberrant behavior; thus, even the gene with a normal allele and a normal eye is termed the eyeless gene.*

1.7 THE GENOME

The genome is an animal's haploid set of chromosomes (*i.e.* 23 chromosomes in a human sperm) and contains all the genes. Double-stranded DNA is usually measured in base pairs rather than nucleotides. The genome contains all the information required to make all the proteins necessary to help make and maintain the operation of the animal. A copy of the entire genome is present in all the cells of an animal that have a nucleus. This means all but the red blood cells and blood platelets of a human have identical genomes. Although the genome consists of nuclear and mitochondrial DNA, the genome term is usually applied to the substantially greater DNA content of the nucleus.

1.7.1 General Features of the Human Genome

The human genome contains about 3 billion (3×10^9) bp in 23 chromosomes, shown in outline in Figure 1.13.

There is a total of about 19 000 protein-coding genes representing only about 2% of the human genome. The function of about 50% of these genes is unknown. The size of the 23 distinct chromosomes range from the largest—chromosome 1 with some 8% of all the genetic information in the human genome and 250 million bp and genes numbering in the thousands—to the smallest male Y chromosome, which is approximately six times shorter, with only 59 million bp and probably less than 100 genes, many involved with the development and fertility of the male. Five acrocentric chromosomes (13, 14, 15, 21 and 22) have very short homologous p-arms with the same structural features and pattern of genes. These contain ribosomal RNA genes and it

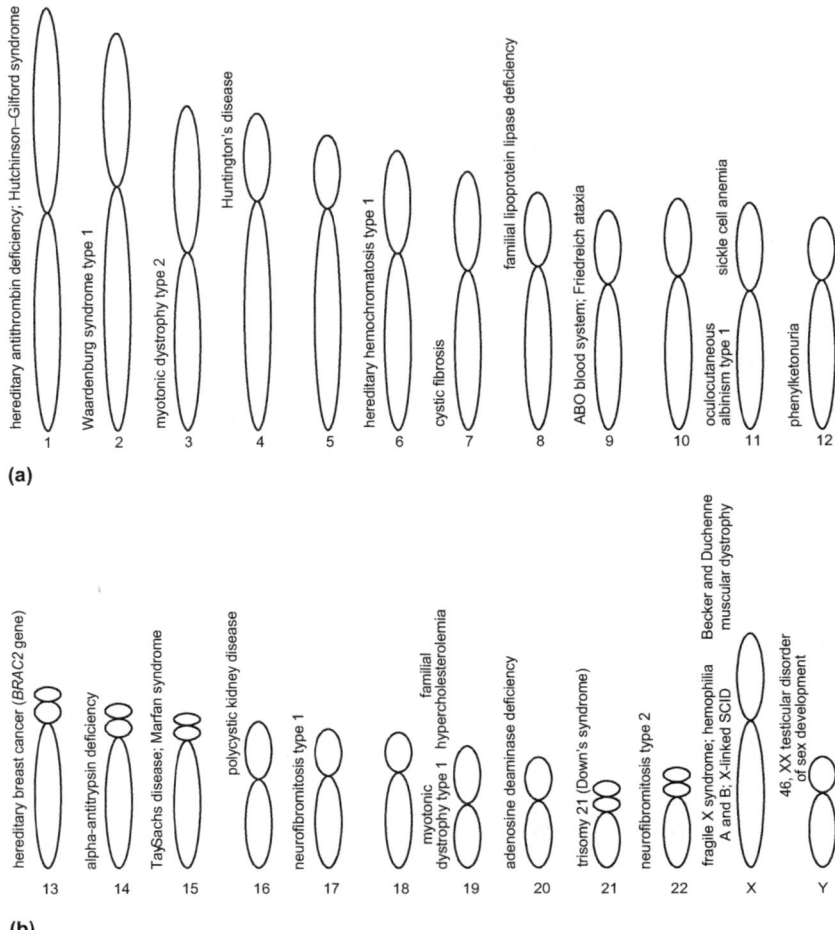

Figure 1.13 The 22 autosomes and two sex chromosomes in humans showing their relative sizes. The location of single-gene-based human diseases considered in the book are shown.

is the q-arms in these chromosomes that contain most, if not all, the protein-coding genes. The gene-dense areas comprise G and C nucleotide pairing, with chromosome 19 being the "gene mother lode". One of the largest "gene deserts" is found on chromosome Y. The gene-poor areas mainly comprise A and T nucleotide pairing. Genes are randomly distributed in humans much more than in other animals. The random areas are stretches of some 30 000 or so CpG clusters called CpG islands, which form a barrier between gene-rich and gene-poor areas. The human genome has a much greater proportion of repeat sequences (50%) compared with other animals (7% in the worm). Humans produce about three times the number of proteins as the fly or the worm, but many of the proteins are common to all three species. The higher production in humans stems largely as a result of alternative splicing (see Section 3.1.4 in Chapter 5).

Diseases discussed in the book that arise from the mutation of a *single* gene are shown in Figure 1.13. Such diseases are relatively rare but *in toto* affect millions of people worldwide and are responsible for a heavy loss of life. They are easier to understand and interpret than the very much more common polygenic diseases in which a number of genes are involved.

1.7.2 Some Features of Interest of 14 Animal Genomes

The comparison of the genome features (DNA sequences, genes, *etc.*) of different animals is termed comparative genomics. Examining regions of similarity and difference can clarify the structure and function of human genes. Knowing the genome of an animal helps to understand some of its unique features and will be helpful knowledge in understanding these features in humans. For example, only about 5% of elephants die from cancer compared with 20% of humans. This is in spite of the much larger number of cells and longer life of the elephant, both factors likely to produce cancer-promoting mutations. Is this because it has 20 copies of the p53 protein tumor suppressor compared with the one copy in humans? Understanding the genetic basis of the longevity of naked mole rats and their almost complete absence of cancer and the fat-processing abilities of the orangutan might be helpful in understanding these diseases and processes in humans. This is possible because a relatively high percentage of genes are shared between all animals, *e.g.*, two-thirds of human genes known to be involved in cancer have fruit fly counterparts. The percentage of human genes with a similar gene in some other animal is shown in Table 1.3.

The values are approximate because authorities differ slightly and genomic length and chromosome division can vary greatly between animals.

Table 1.3   Percentage of genes shared by humans with other animals.

Cat, 90	Chicken, 60	Chimpanzee, 98
Cow, 80	Dog, 83	Fruit fly, 46
Honeybee, 44	Mouse, 84	Platypus, 82
Rhesus monkey, 93	Roundworm, 38	Zebrafish, 76

It was originally believed that the additional number of genes in humans than, for example, in chimpanzees would explain their "superiority". It now appears that it is the loss of certain segments of DNA that may explain human dominance. For example, there is a gene in animals that stifles the growth of brain cells. Since it is missing in humans, this might explain their bigger brain. Some 20 or more mammals, with many others on the way, as well as other animals, plants and microbes, have had their genomes measured, mainly using shotgun sequencing (see Section 2.1.4 in Chapter 2). Features of interest of a number of animals are briefly described below.

- **Honeybee.** Compared with the fruit fly and mosquito, the honeybee has twice as many genes coding for odor-sensing proteins, which are essential for recognizing flower types. However, it has fewer genes related to detoxification, possibly making them more sensitive to environmental insecticides, in part explaining the recent devastation of honeybee colonies.
- **Monarch butterfly.** Genetic adaptations involving brain, antennae and eyes allow them to survive their long (approximately 4000 km) journeys. Migrating monarchs lack the gene that encodes a key enzyme that produces a hormone known to stimulate the reproductive organs. This might mean they are disinterested in sex and can focus on their long flights!
- **Chicken.** Unlike mammals, the chicken genome contains a large number of microchromosomes. A particular gene is common only to the chicken and human. This gene features in the immune response so that, for example, avian flu can move easily between the chicken and human (where it can be deadly). The chicken and turkey genomes bear a close resemblance.

- **Cow.** Perhaps surprisingly, the human genome more closely resembles that of the cow than that of rodents. There is high genetic diversity, particularly within the Hereford breed. More genes are linked to the immune function in cows than in humans and it is suggested that this is needed in cows due to their multiple stomachs, which harbor more micro-organisms. Cattle breeders are mapping single nucleotide polymorphisms (SNPs; see Section 1.5.1) in their animals in order to identify which SNPs are linked with higher milk quality and better beef in the animal.
- **Dog.** Although domestic dogs vary widely in appearance, their genomes are 99.85% similar. For example, the boxer and the poodle differ by only one single nucleotide in about every 900 bases. They have more genes in common with humans than mice and are therefore helpful in studying common diseases like diabetes, cancer, *etc.* Their genomes indicate that about 75% of all dogs now on Earth descended from a particular family of wolves around 40 000–100 000 years ago.
- **Elephant.** Both nuclear and mitochondrial DNA analyses indicate that the Asian elephant is the closest living relative to the extinct woolly mammoth, although they are considered to be distinct genera. The Savanna and smaller African forest elephants have even wider divergent genomes and should be considered different species, contrary to previous beliefs (see Section 1.7.3).
- **Housefly.** The genome of the housefly, very recently sequenced, contains 691 Mbp compared with the *Drosophila melanogaster* fruit fly (123 Mbp). It carries many pathogens (disease-carrying microorganisms) to which it is immune but can transmit to humans and animals, possibly leading to over 100 diseases. Not surprisingly, it has many diverse immune genes that help protect it against these pathogens. This is the type of knowledge that could help in the understanding of a number of human and animal diseases.
- **Rhesus monkey.** This was the first primate cloned and its genome determined. The monkey is an important primate model for the study of human diseases because of the genetic, as well as physiological and metabolic, similarities to humans. It is commonly used for drug testing. Genetic

analysis, as well as other evidence, indicates that rhesus monkey ancestors diverged from humans 25 million years ago and from chimpanzees 6 million years ago.

- **Pig.** There are thousands of pig genes or proteins that are very similar in their nucleotide and amino acid sequences (orthologs) to those found in other mammals. Genome information indicates that pig and wild boar populations diverged about 1 million years ago and the pig was domesticated about 10 000 years ago. The pig is an important model for human health.

- **Platypus.** It was inevitable that such an animal with strange features, such as a wide and flat tail like a beaver, a bill and webbed feet like a duck, and which lays eggs like a bird, should have its genome examined. It turns out that its genome is an eclectic brew of bird, mammal and reptile genomes. Examining the platypus genome might give clues as to the function of certain mammalian genes. For example, understanding the gene responsible for testicular (one copy in platypus, two copies in humans) descent may help the 30% of prematurely born boys who have testes that fail to descend properly.

- **Hairless naked mole rat.** This is the only cold-blooded mammal that lives in the dark. Several genes related to vision are turned off. Searching for significantly different genes from those of humans and mice may help explain their longevity (about 30 years, ten times longer than other similar-sized rodents) and resistance to some diseases, such as cancer (the naked mole rat does not develop cancer even when cancer cells are implanted in its body).

- **Sea urchin.** This creature has a large toolbox of genes that combat infection, which part explains its approximate 100-year life span. The animal has been used to study fertilization and early development in humans, in which there are similarities.

- **Tammar wallaby.** This is a member of the kangaroo family. A female has been sequenced using the shotgun approach and Sanger sequencing (see Section 2.1 in Chapter 2). The genes in the kangaroo ($2n = 16$, *i.e.*, 16 total chromosomes) are similar in large part to normal placental mammals, but it has a small number of very large chromosomes.

- **Tiger.** The endangered Siberian tiger has had its genome compared with that of the Bengal tiger, African lion and snow leopard. Overall, the cat family, including the domestic cat, relies on a narrow set of 1376 genes (!) linked to strong muscle fibers and the ability to digest proteins (meat), both prerequisites for a hunter. DNA recovered from tiger skins, which were hunting trophies from a century ago, is much more variant than that of today's tiger. This decrease in genetic diversity over this time is a recognized signal for a dangerous direction towards extinction.

1.7.3 Aged Specimens

Very old DNA from the mastodon (50 000 years old), polar bear (110 000 years old), horse (700 000 years old), wolf-like dogs (30 000 years old) and wooly mammoth (20 000 years old) has provided partial or complete genomic information. These animals had been preserved in cold storage provided by the permafrost. Contaminating DNA from bacteria and fungi needed to be removed before analysis.

- **Dog.** By comparing mitochondrial DNA of domestic dogs and ancient wolf-like and dog-like animals (some specimens were 30 000 years old), it was surmised that domestic dogs are genetically grouped with ancient wolves or dogs from Europe and not with modern European wolves.
- **Woolly mammoth.** Large amounts of the nuclear genome have been deciphered using DNA from hairs or bone of a mammoth mummy buried in the Siberian permafrost for 20 000 years. Combined with data for the entire mitochondrial genome sequence, this suggested that the mammoth and modern-day elephants separated about 6 million years ago. Using this gene information, it has been possible to reconstruct the hemoglobin used by the woolly mammoth and to rationalize the ability of the animal to withstand the extreme cold it experienced. Fourteen of the mammoth's genes have been inserted into the DNA of a live elephant using the CRISPR/Cas9 technique (see Section 8.4.10 in Chapter 8) for a precise replacement. These genes were associated with cold-resistant properties missing from the

elephant, such as hairiness, ear size and its hemoglobin. The inserted genes functioned normally. This is an important step to the controversial recreating of a woolly mammoth.

- **Domestic cattle.** DNA from bones of domestic cattle from about 10 000 years ago, when cattle were first domesticated, was compared with DNA from present-day cattle, which suggested that all cattle are descended from as few as 80 animals domesticated from wild ox.
- **Neanderthal.** These existed from several hundred thousand years until about thirty thousand years ago, when they became extinct. An entire genome of high quality has been obtained using the toe bone of a 130 000-year-old Neanderthal found in a Siberian cave. Indigenous Africans and Neanderthals have little or no DNA in common, consistent with Africans not breeding with Neanderthals who lived in Europe and Asia. About 30 000 years ago, *Homo sapiens* migrated from Africa and began encountering Neanderthals. About 2–4% of the genes across the genome of the Neanderthal are found in humans of wide origins, except Africans. The number of mutations common to the two species is relatively small and it is suggested that harmful mutations in the Neanderthal were eliminated by natural selection. Recently, the jawbone of a 35 000-year-old "Romanian" man with features common with those of a Neanderthal shared as much as 6 to 9% of the genome and analysis of their DNA suggested that the man had a Neanderthal ancestor only 4–6 generations earlier.

1.7.4 The Mitochondrial Genome

Because of the very much smaller DNA content of the mitochondrion compared with that in the nucleus, the determination of the mitochondrial genome is a much less difficult task and a large number of different animals have provided mitochondrial genome information.

- Each human mitochondrion contains 5–10 copies of its genome. Since there are 50 to hundreds of mitochondria depending on the cell type, this means 10^3–10^4 copies per cell. These copies are usually identical in a cell in a healthy person. However, low levels of heteroplasmy may exist, in which more than one mitochondrial DNA haplotype (haploid genotype) coexists within a cell or in different tissues of an animal (see Section 6.4.2 in Chapter 6 and Section 8.1.3 in Chapter 8 for further elaboration).

- Human mtDNA is a 16 500 bp circle (Figure 1.14). The arrangement of genes in mtDNA resemble those in bacteria more than those in nDNA.

- With few exceptions, all animals contain 37 genes encoding two ribosomal RNAs (rRNAs) and 22 transfer RNAs (tRNAs) used in translating mRNA into polypeptides. There are also

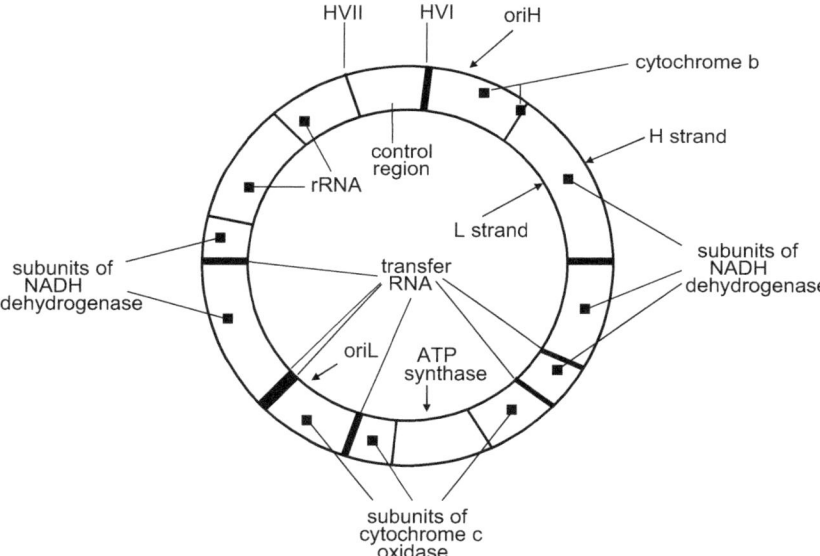

Figure 1.14 The human mitochondrion genome showing some of the main features. The approximate positions of the 13 protein-encoding genes are shown. These are (number in brackets) cytochrome b apoenzyme (1), subunits of NADH dehydrogenase (7), cytochrome c oxidase (3) and ATP synthase (2). Also shown is the location of the small and large ribosomal rRNAs. The thickened black lines represent the gene regions for 22 tRNAs (six are named), which include the 20 amino-acid-specific transfer RNAs (see Section 3.1.1 in Chapter 3). Very similar features apply to the mitochondria of all animals.

13 genes encoding proteins that are involved in synthesizing ATP, which is important in oxidative phosphorylation, a major task of the mitochondria. The heavy strand (H) is nucleotide G rich and contains the majority of the genes (28 in humans). The light strand (L) is nucleotide C rich and, in humans, contains nine genes.

- All genes lack introns and there are very few or no intergenic non-coding sequences (see Section 3.1.3 in Chapter 3). Sometimes, genes overlap so that nearly every base pair is part of a coding gene.

Figure 1.14 shows the main features of the human mito-chondrion genome (the word genome is often omitted). The two regions, hypervariable HVI and HVII, are often used for the identification of human remains that have highly degraded DNA and in forensic investigations (see Section 8.1 in Chapter 8). The *cytochrome b* gene is most commonly used for species identification and phylogenetic studies because it shows high interspecies conservation but is sufficiently variable to allow interspecies differentiation. Mitochondrial DNA is used for mapping ancestry and migration patterns. The study of mito-chondrion genomes and the comparison of mitochondrial gene arrangements are useful for genome evolution and ancient evolution relationships since mutations that might spoil the interpretation are rare events.

A now classic example of its use is in the confirmation of the skeletal remains as Richard III (see Section 6.4.1 in Chapter 6).

CHAPTER 2

The Exogenous Manipulation of DNA

2.1 DNA SEQUENCING

2.1.1 The Chain Termination Method (Sanger Sequencing)

The procedure is elegant and relatively straightforward, and is shown in Figure 2.1 and outlined below.

- A number of single-stranded target DNA fragments are generated by heating the double-stranded DNA.
- These are mixed with a short primer and four different deoxyribonucleotide triphosphates, which are collectively termed deoxynucleotides (dNTPs), namely deoxyadenosine triphosphate (dATP), deoxycytidine triphosphate (dCTP), deoxyguanosine triphosphate (dGTP) and deoxythymidine triphosphate (dTTP). Also added are relatively small amounts of the four dideoxynucleotide triphosphates, ddNTPs (Figure 1.6, Structure B), namely ddATP, ddCTP, ddGTP and ddTTP, each with a different fluorophore label. This means they will fluoresce with different colors, *e.g.*, green, yellow, blue or red, when irradiated with UV light and can thus be identified. Like dNTP, ddNTP will also add to DNA, but since it contains no $3'$–OH group, once added this will terminate the building of the chain (Figure 1.7; X = Y = H). The single,

Animal Genetics for Chemists
By Ralph G. Wilkins
© Ralph G. Wilkins 2017
Published by the Royal Society of Chemistry, www.rsc.org

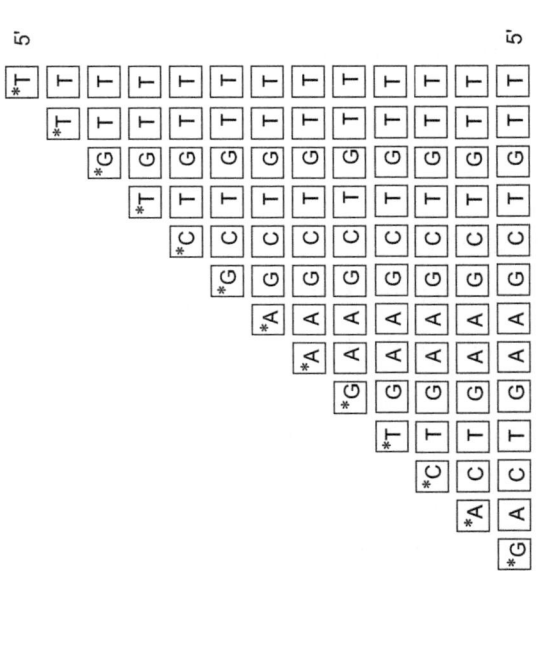

Here, the sequence of the DNA strand must be:

5' C T G A C T T C G A C A A 3'

Figure 2.1   Sanger sequencing. Each ddNTP is labeled with a selective fluorescent "tag" indicated by an asterisk and terminates the 13 different fragments.

short primer has a nucleotide sequence that is complementary to the 3′ end of the region to be copied and
therefore defines it (see Section 2.3.2 also). The primer is
also necessary to provide a short double strand for DNA
polymerase to initiate DNA replication (see Section 3.3.1 in
Chapter 3).

• Addition of DNA polymerase starts the process. Chain
elongation proceeds normally by the addition of the appropriate dNTP (Figure 1.7) until, by chance, DNA polymerase
inserts the appropriate ddNTP and termination results. Some
DNA strands may add no dNTP before termination by the
addition of ddNTP (*T in Figure 2.1); some other strands may
add 12 dNTPs before termination (by *G in Figure 2.1) or even
hundreds of dNTPs before a ddNTP is added by chance and
the chain is terminated. What results, therefore, is a series of
fragments of different sizes, terminating at a ddNTP—the
nature of which by its generated fluorescent color tells us the
identity of the nucleotide to which it became attached.

Capillary Gel Electrophoresis

The characterization of this mixture is now possible by capillary
gel electrophoresis using powerful automation techniques.
A series of sequencing products is each loaded into a 96-well
cathode plate and each well is connected to an ultra-thin
(0.1 mm diameter) capillary tube that is 50–80 cm long and filled
with a resin bead mixture. Application of a high voltage causes
the (negatively charged) DNA fragments in the well to enter the
capillary and migrate to an anode reservoir attached to the other
end of the capillary. Chain-terminated fragments migrate
through the gel, the smaller fragments (containing the bases at
the beginning of the DNA sequence) moving faster. At a position
near the end of the capillary a laser activates the particular
ddNTP color and a photometer records the color and therefore
the nucleotide responsible. Thus, a procession of fragments, the
smaller ones earlier, pass by the laser point and produce an array
of colored bands, which identify the chain-terminating nucleotides. The sequence of the DNA can thus be recorded as a series
of colored peaks (electropherogram), with each different color
representing a different nucleotide. With an automated machine

using as many as 96 capillaries operating simultaneously it is possible to determine about 100 different sequences in a 2-hour period, assuming a maximum of about 750 bp sequenced in a single experiment. Differences of one nucleotide can be detected by the method. It is good for small-scale sequencing, but there is poor quality for the first 15–40 bases because of interference by the primer binding and deterioration in the accuracy of the sequencing after 700–900 bp have been sequenced.

2.1.2 Pyrosequencing

This method uses a quite different principle than that used in the Sanger method. In some respects, it is simpler since it does not require the use of dideoxynucleotides nor gel electrophoresis. It is based on the fact that binding of a nucleotide to DNA releases pyrophosphate (PP_i; Figure 1.7). Thus, the method requires the detection of pyrophosphate and this is achieved by a clever use of a sequence of reactions in which pyrophosphate is converted into a pulse of light *via* the generation of ATP. The chemistry involved is shown in Figure 2.2A.

A stretch of DNA that is to be sequenced is denatured (in alkali) to single strands, amplified by PCR (see Section 2.3), a sequencing primer attached and the whole incubated with the enzymes DNA polymerase (catalyzes the chain building), ATP sulfurylase, luciferase and apyrase, as well as adenosine $5'$ phosphosulfate (APS) and luciferin (Figure 2.2). The four dNTPs are added sequentially and repetitively to the singly stranded DNA target using an automated machine. Only when the correct one of the four deoxynucleotides binds to the DNA by hybridization is pyrophosphate released and this is accompanied by a light pulse, Figure 2.2A. After the dNTP addition, any dNTP that is not incorporated in the DNA chain, and any leftover ATP, are rapidly degraded enzymatically before the next nucleotide is added. This uses apyrase, which is an ATP diphosphohydrolase, which removes phosphate groups (P_i) from dNTP and ATP (Figure 2.2B). The result of the procedure is a pyrogram, Figure 2.3.

An examination of Figure 1.7 shows that hydrogen ions, as well as pyrophosphate ions, are released in the binding of a nucleotide to the DNA strand. The detection of the hydrogen ion as a voltage change replaces the pyrophosphate ion as a light

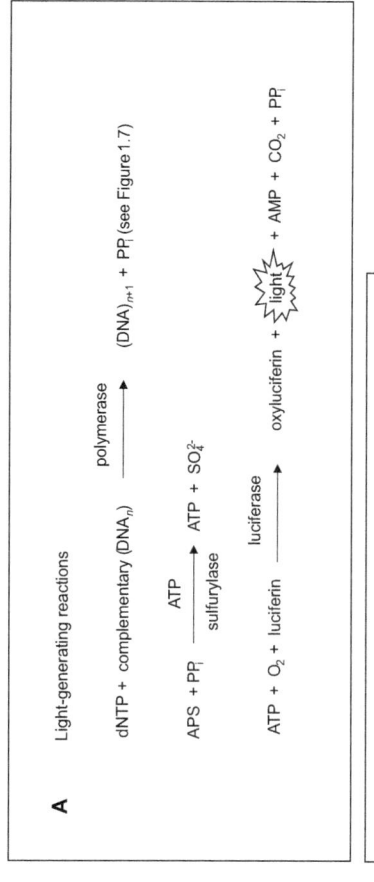

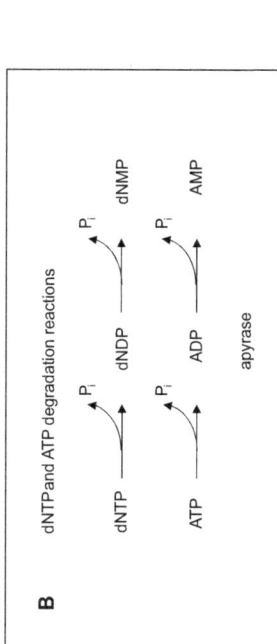

Figure 2.2 (A) The conversion of pyrophosphate (PPi) into a light signal. APS = adenosine 5'-phosphosulphate. (B) The degradation of unincorporated dNTP and excess ATP using apyrase (AMP, adenosine monophosphate; ADP, adenosine diphosphate; P$_i$, phosphate).

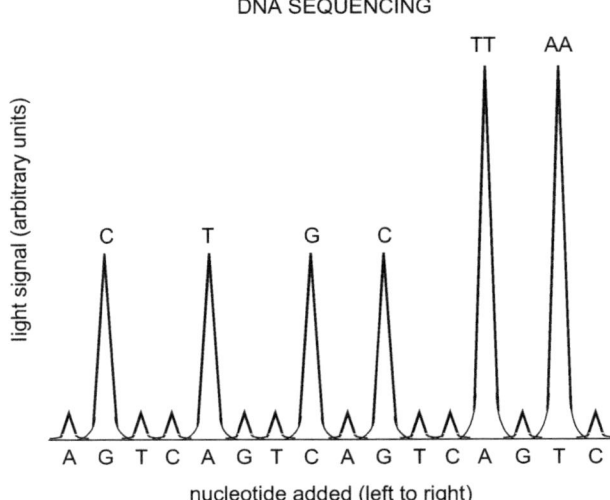

DNA SEQUENCING

Figure 2.3   A pyrogram. Continual, repetitive succession of A, G, T and C
nucleotides are added to the target DNA. The first addition (A)
does not bind and no light is emitted. The first pulse of light is
recorded when the next added nucleotide (G) reaches and binds to
the DNA (C). The next nucleotide additions (T and C) do not bind
(no light emitted), but the following nucleotide (A) does, and so the
next nucleotide on the target DNA must be T. At one point, a light
pulse of double the usual intensity is observed when A is added,
indicating that two adjacent T nucleotides exist on the target DNA.
In this way, the whole sequence is obtained. Therefore, in this
case, the sequence is CTGCTTAA.

pulse, but otherwise the procedure is exactly the same in this
variation of DNA pyrosequencing.

2.1.3   Bisulfite Sequencing

In epigenetics studies, it is important to know the state of
methylation of DNA and specifically of the cytosine component.
Bisulfite sequencing is the gold standard for DNA methylation
analysis. Bisulfite removes the $-NH_2$ group in cytosine and
converts it to uracil. It does not, however, react with methylated
cytosine (Figure 2.4).

The DNA is sequenced before and after bisulfite treatment.
The only cytosine that remains as cytosine after bisulfite treatment must have been methylated (Figure 2.5).

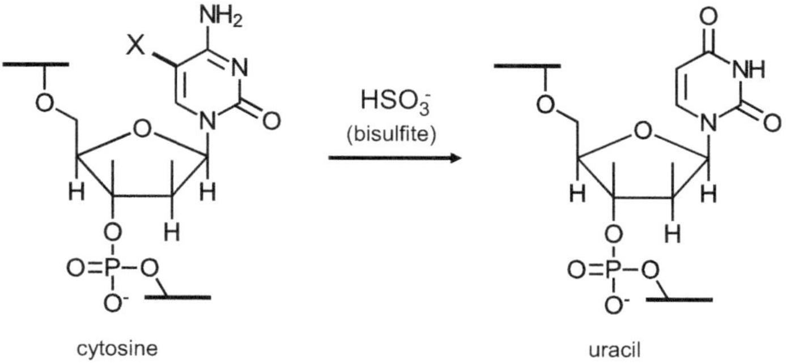

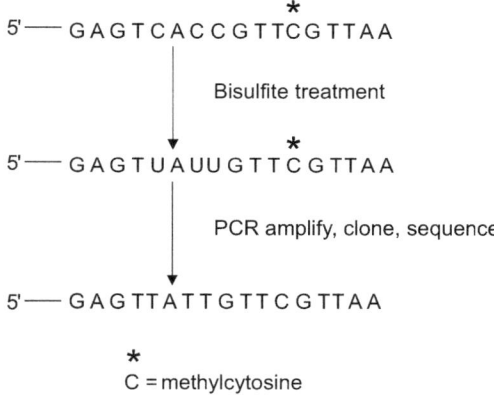

Figure 2.4 The bisulfite conversion of cytosine (X = H) to uracil in a DNA fragment. The reaction proceeds with cytosine (X = H) but not with 5-methylcytosine (X = CH_3).

5'——G A G T C A C C G T T*CG T T A A

Bisulfite treatment

5'——G A G T U A UU G T T C*G T T A A

PCR amplify, clone, sequence

5'——G A G TT A T T G T T C G T T A A

*
C = methylcytosine

Figure 2.5 Single-stranded DNA from the carefully denatured DNA before and after bisulfite treatment. A methylated C (asterisk) is not affected by bisulfite, whereas unmethylated cytosine C is converted to uracil. 5-Methylcytosine is amplified as cytosine, and uracil as thymine by PCR in the sulfite-treated DNA, which is then sequenced and compared with the original. The methylation patterns of both strands of the DNA can be assessed.

2.1.4 Next Generation Sequencing

First generation sequencing is embodied in Sanger sequencing and took center stage from the mid-1970s until the early 2000s, with the completion of the first human genome sequence. It is still in use but has been augmented and, in some cases, replaced by the next generation sequencing (NGS) technique, which is also termed massively parallel sequencing. In this, thousands to millions of fragments of DNA cut from a single, large DNA segment are sequenced in unison. Several variations, termed platforms, have been developed, but the basic four steps in the procedure are common to all.

1. DNA from different sources, perhaps genomic DNA, is randomly cut by mechanical shearing, *e.g.*, by sonication or by random enzymatic digestion, to give short, single-stranded DNA fragments, around 300–500 bp in length.
2. Clone amplification of each fragment is effected by emulsion PCR (see Section 2.3.3) so that enough material is available for subsequent sequencing. The several hundred thousand wells, each containing one amplified fragment resulting from this treatment, are packed with the necessary reagents for the chosen sequencing method.
3. The thousands to millions of amplified and segregated DNA templates are now sequenced simultaneously in a massively parallel fashion. Sequencing is usually by synthesis rather than through the chain-termination approach. Commercial equipment is available that uses a number of strategies, including pyrophosphate and proton detection.
4. Shotgun sequencing. The task now is to match up many fragments so as to give the desired contiguous sequence. This is much like assembling a jigsaw puzzle but where the various pieces only partially fit! A very simplified idea of the approach to this problem is shown in Figure 2.6.

Powerful computers are essential in the correct assembling and the process is repeated a number of times to improve the accuracy of the procedure. The most recent machines can provide 1 million reads occurring simultaneously, with an average read of 400 bases. A 10-hour run time can provide 400 million bases of sequence information. The method is complex but rapid. The sequencing of the large, complete DNA (genome) of an animal is obviously a sizeable problem, but in the last decade or so this issue has been solved successfully using NGS. In addition, it is becoming more affordable. Such sequencing at a

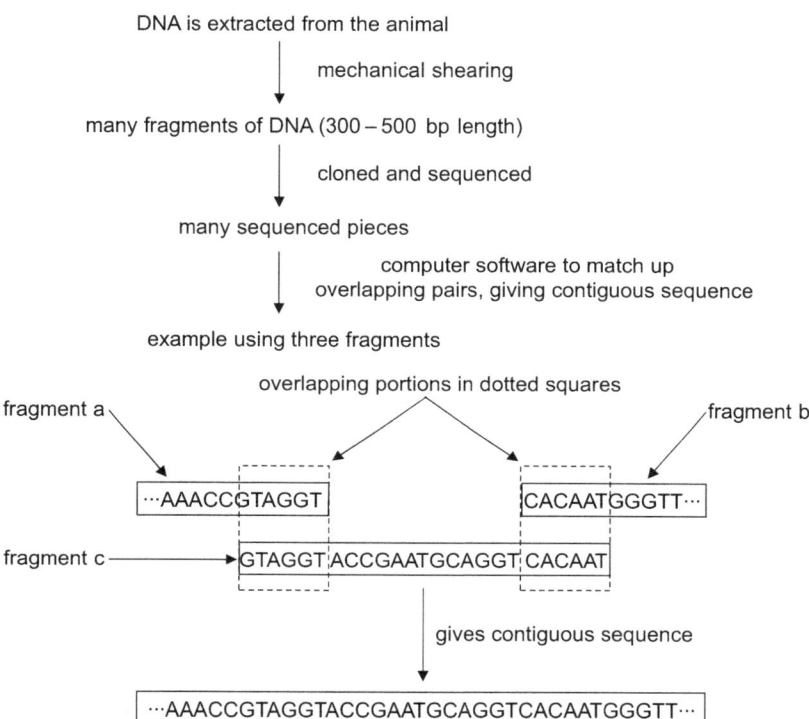

Figure 2.6 Summary of the process for the sequencing of an animal genome, much simplified.

single cell level may cost only $1000 or so. A number of animal genomes (or substantial portions thereof) have been determined (see Section 1.7.2 in Chapter 1). Massively parallel pyrosequencing has been used in sequencing, for example, the genomes of the woolly mammoth, Neanderthal man and the DNA pioneer James Watson. It is very effective for sequencing mitochondrial DNA and for studying the human microbiome, *i.e.*, the microorganisms in the body. Sometimes, it is unnecessary to break up the sample DNA in the first step, which may survive in fossils as short fragment DNA, and from this the full genome sequence can be obtained. This has been possible in the sequencing of the mastodon (50 000 years old), polar bear (110 000 years old) and the Yukon horse (700 000 years old), all of which were preserved in cold storage in the permafrost (see Section 1.7.2 in Chapter 1).

Nanopore Sequencing

The simplest method of sequencing in principle, which avoids the complex methods currently used, is slowly reaching fruition. Single DNA is passed through a protein nanopore approximately 1 nm in diameter and each of the four DNA bases are specifically detected by an electrochemical method. Long reads may be achieved at high speed (450 bases per second).

2.2   GENE CLONING

Gene cloning is the production of an exact copy of a single gene of an animal's genome by propagating this copy in bacteria. There are various ways to carry this out. A vector (see Section 2.2.1) may be used to carry the segment of DNA that needs to be cloned into a host cell, where it is replicated. The transfer into the host cell may be carried out without using a vector (although this is less popular). The use of both a vector and a cell may be dispensed with altogether.

2.2.1   Gene Cloning with Vectors

The gene is inserted into a vector carrier, which is then encouraged to multiply in a host cell (usually bacteria) in a laboratory to give many copies of that gene.

Vectors

- In bacteria, plasmids are double-stranded, circular DNA, which are found in bacterial cells but rarely in animal cells. Plasmid DNA is separate from the normal DNA of the containing cell and is capable of replicating independently of the cell DNA. The plasmid vector is made from the natural plasmid by removing unnecessary segments and adding essential sequences. Its length is shortened to approximately 3 kb. Most commonly, plasmids are replicated in *Escherichia coli* (*E. coli*). The limit of the amount of DNA that plasmids may clone is 0.1–10 kb, but a modification (cosmid) can extend the cloning limits to 35–50 kb. One of the most popular *E. coli* cloning vectors is the plasmid pUC8 (Figure 2.7).

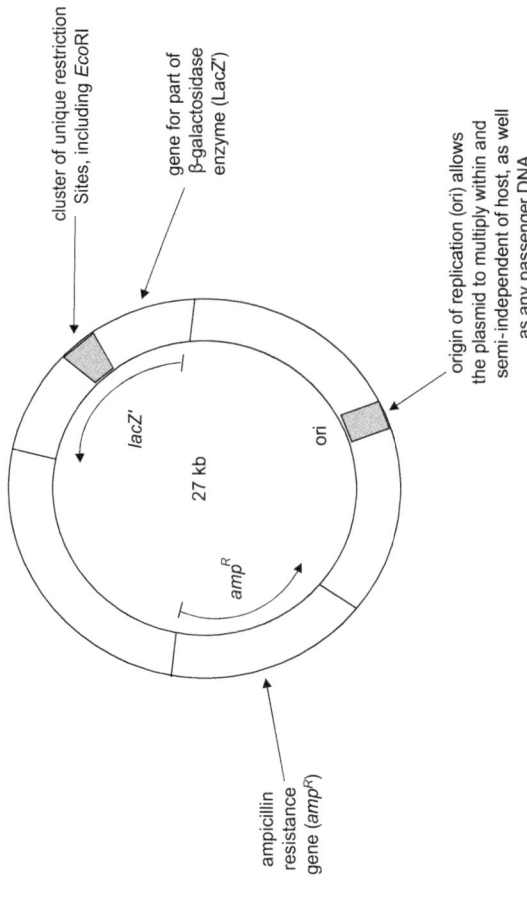

cluster of unique restriction
Sites, including *EcoRI*

gene for part of
β-galactosidase
enzyme (LacZ′)

origin of replication (ori) allows
the plasmid to multiply within and
semi-independent of host, as well
as any passenger DNA

lacZ′

ori

27 kb

amp^R

ampicillin
resistance
gene (*amp^R*)

Figure 2.7 Characteristics of a vector. The vector pUC8 contains four relevant sites. These are the origin of replication (ori), restriction sites (where restriction enzymes operate), an ampicillin-resistant gene (amp^R) and a *lacZ′* gene, which encodes for part of beta-galactosidase, an enzyme that helps break down lactose into glucose and galactose.

- Viruses can infect cells and therefore inject their DNA into host cells. Viruses are normally altered so that they have no side effects and can act as vectors. They are more efficient than plasmids. One virus vector that we shall meet in gene therapy (see Section 8.4.5 in Chapter 8) is the adeno-associated virus (AAV), a single-stranded DNA that can insert genetic material at a specific site on human chromosome 19.

Restriction Enzymes

Restriction enzymes (endonucleases) are used to splice DNA into smaller fragments at specific nucleotide sequences. There are many restriction enzymes of different types. In our example, *E. coli* RI (*Eco*RI) cuts DNA between bases G and A only when the sequence GAATTC is present in the DNA to give a DNA fragment with overhangs ("sticky ends"). This can anneal (hybridize) to any DNA fragment produced by that same *Eco*RI restriction enzyme (Figure 2.8).

The overall process for gene cloning is shown in Figure 2.9.

The DNA is cleaved into a number of fragments of different sizes but with the same "sticky ends" controlled by the restriction enzyme *Eco*RI. One of these fragments will contain the gene that we are wishing to clone. The same "sticky ends" have been created on the plasmid pUC8, which has been opened up at the single recognition site by *Eco*RI (Figure 2.9). The various fragments of DNA will fit nicely into the opening, where they can be joined and sealed (by the enzyme ligase). We now have to consider the fate of several entities arising in the process. These include:

- The original unligated DNA fragments and opened vector molecules that are unligated. These will be most likely enzymatically degraded in the host bacteria. More important we also have:
  - unchanged, non-recombinant plasmid, which has self-ligated without incorporation of foreign DNA,
  - recombinant plasmids with DNA fragments of different sizes but not containing the desired gene and
  - desired recombinant plasmids containing the gene of interest.

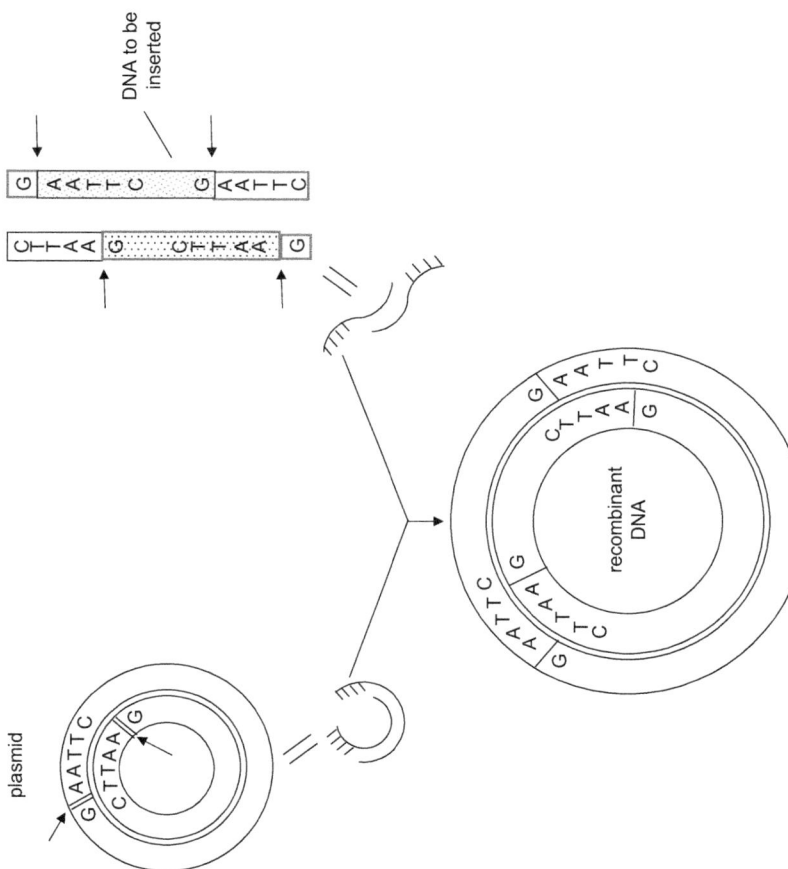

Figure 2.8 Attachment of one DNA fragment to another, in this case using the restriction enzyme *EcoRI* for incorporation of a DNA fragment into a plasmid. The DNA is cut at the arrows using *EcoRI* and the sticky ends joined using DNA ligase.

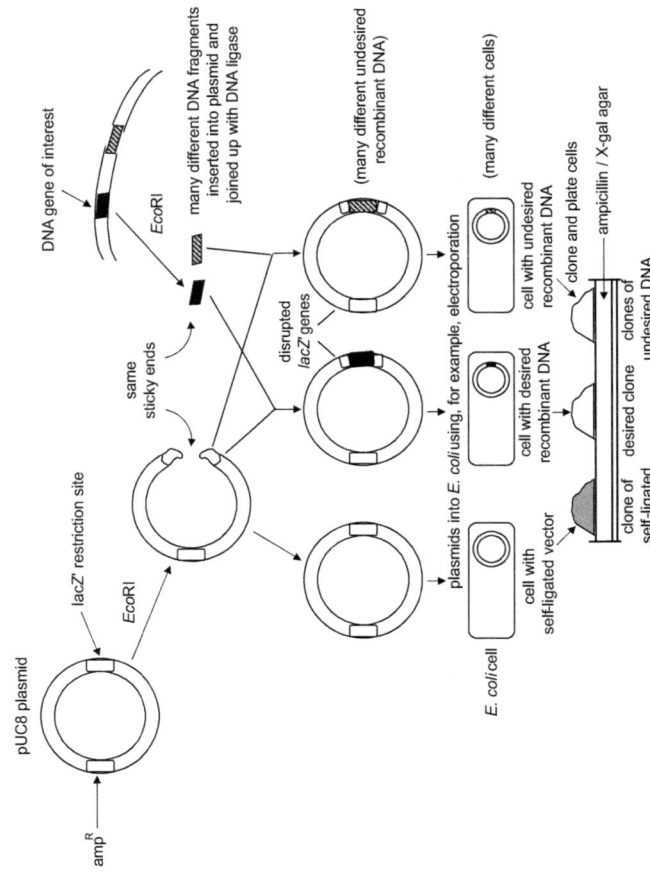

Figure 2.9 Vector gene cloning producing bacterial clones with many copies of a number of different recombinant plasmids.

All these plasmid vectors are introduced (transformed) into the host *E. coli* cell, with their entry aided by chemicals, such as calcium chloride, or by electroporation (see Section 2.2.2).

- The plasmids multiply by replication independently of the DNA of the host cell. After a large number of cell divisions, colonies or clones of identical host cells (but of many compositions) are obtained. How can we differentiate between these? We use a modified form of *E. coli*.

LacZ *Gene*

The natural *E. coli* has a *LacZ* gene, which synthesizes a vital enzyme, β-galactosidase, which is detectable by conversion of a lactose analog (X-gal) to a blue coloration. A modified *E. coli* strain that we use has a non-functional gene, with a large segment (*LacZ'*) of the normal *LacZ* gene missing. With this strain, X-gal is not converted to a blue color. When the plasmid is incorporated into the modified *E. coli* strain it can supply the missing segment *LacZ'* but only provided the relevant site is not disrupted by any foreign DNA inclusion there. The plasmid also contains a gene that confers antibiotic (ampicillin) resistance on the otherwise antibiotic-sensitive host cell.

- The modified *E. coli* is grown on an agar plate that contains ampicillin and X-gal. The ampicillin kills bacterial cells without plasmid. Colonies carrying non-recombinant plasmid will have the blue color because they have an intact *LacZ* gene and will grow because they have the ampicillin gene. Colonies carrying the recombinant plasmid with the disrupted *LacZ* gene will be white but will grow because of the intact ampicillin-resistant gene (Figure 2.9).
- The final and most difficult stage in many respects is the assignment of the cell clone to the required gene (screening). The general technique is shown in Figure 2.10.

There will be many different clones with different lengths of DNA and different DNA sequences, all of which have the identical termini required by the *Eco*RI enzyme. If the DNA sequence of the gene of interest or the encoded protein is known, the desired identification is eased. In this case, in one method the

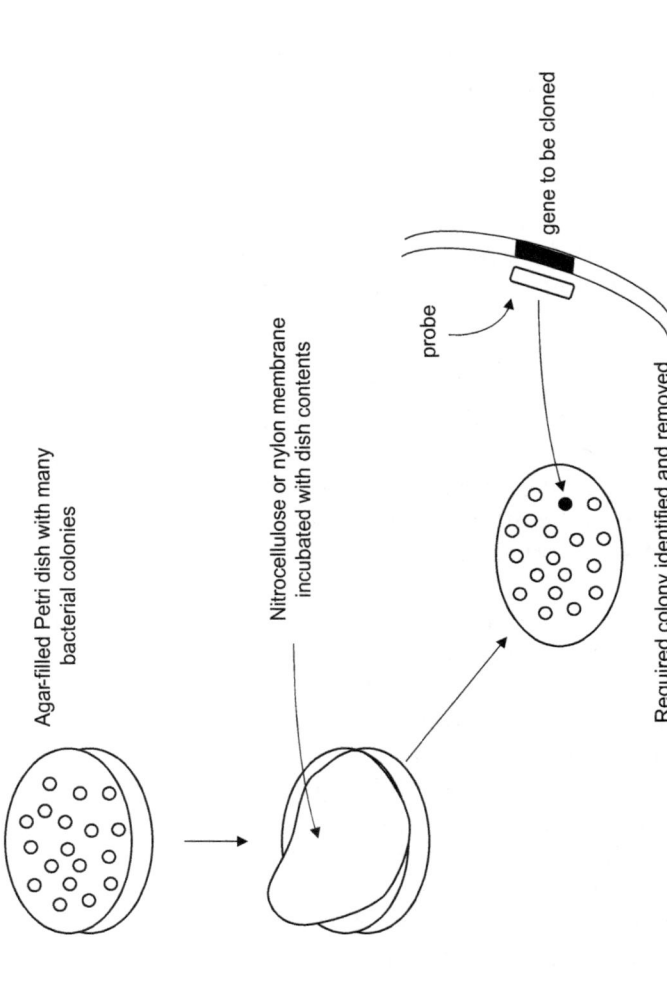

Figure 2.10 Screening for the colony containing the required gene. Incubation of the membrane with the disc contents attaches enough of the denatured cloned DNA segments in the colonies to the membrane. The lifted membrane is treated with a fluorescent probe, which is designed to attach to the sought gene by hybridization, and thus its position on the plate determined.

DNA strands in the plasmid are denatured. The colony (one of many) containing the desired gene with a unique DNA sequence is then found using a probe. This has a complementary DNA sequence to that of the gene. It is tagged by radioactive or fluorescent labeling so that its position can be found and the required colony on the master plate thus identified. It can be picked off the master plate with a thin needle and allowed to multiply.

Recombinant Protein Production

In the simpler case of only cloning a single DNA segment, *e.g.*, a gene not part of a large DNA fragment like above, the final steps are obviously unnecessary. The bacteria containing only the one required type of recombinant plasmid is lysed and the hybrid plasmids isolated. Cutting these with *Eco*RI will release many copies of the DNA fragment. Alternatively, the bacterial system that includes the necessary ribosomes can be induced to generate the encoded protein. This is the basis of the 1980s production of human insulin. The human insulin-producing gene from the pancreas cell is designed to have sticky ends, which will enable it to be incorporated into plasmid DNA. This entity in *E. coli* makes human insulin, which can be extracted and purified.

2.2.2 Gene Cloning Without Vectors

Effecting *direct* entry of the DNA segment into the host cell, which usually does not work, can be aided by mechanical or chemical means:

- Temporary pores produced by a high voltage may permit the entry of foreign DNA (electroporation).
- Certain chemicals, such as calcium chloride, sometimes help the entry of foreign DNA into a host cell.
- Microinjection of linear DNA copies of a gene into mammalian nuclei results in a tandem head-to-tail arrangement of DNA in the chromosome.
- Mechanical delivery of DNA on microscopic particles (tungsten-coated) with a gun (biolistics) has been used mainly in plant genetics but is increasingly being considered for the production of transgenic animals.

The use of vectors and cells for cloning purposes can be avoided completely by using PCR.

2.3 THE POLYMERASE CHAIN REACTION (PCR)

The goal of PCR is to make millions or billions of copies of a specific part of DNA. Its value in the studies of molecular biology and animal genetics cannot be over-emphasized. It is frequently employed to amplify DNA fragments in genes for analysis. It is also much used in forensics. Many copies might be necessary for DNA in blood or semen taken from a crime scene. Amplified DNA may be needed from samples as diverse as single embryonic cells for prenatal diagnosis or from frozen ancient mammals.

2.3.1 The PCR Process

The process is shown in Figure 2.11.
    The components required are:

- A selected region of DNA, usually less than 3 kb but ideally less than 1 kb in length.
- Taq DNA polymerase from *Thermus aquaticus*, which is present in hot springs.
- A plentiful supply of nucleotides.
- A pair of primers.

2.3.2 Primer Design

Two non-identical primers [oligonucleotides (see Section 1.2.8 in Chapter 1) about 18–22 bases in length] are annealed to opposite strands of the DNA template (Figure 2.12).
    Their sequences must be complementary to the sequences of the flanking target region on the template molecule. The 3' ends of the primers point at each other. The forward and reverse primers should not even be partially complementary, otherwise they will hybridize to form a primer dimer. This does not usually happen because of the length of the primer employed. The primer thus defines the ends of the DNA segment that is to be duplicated. The primers are also necessary because DNA

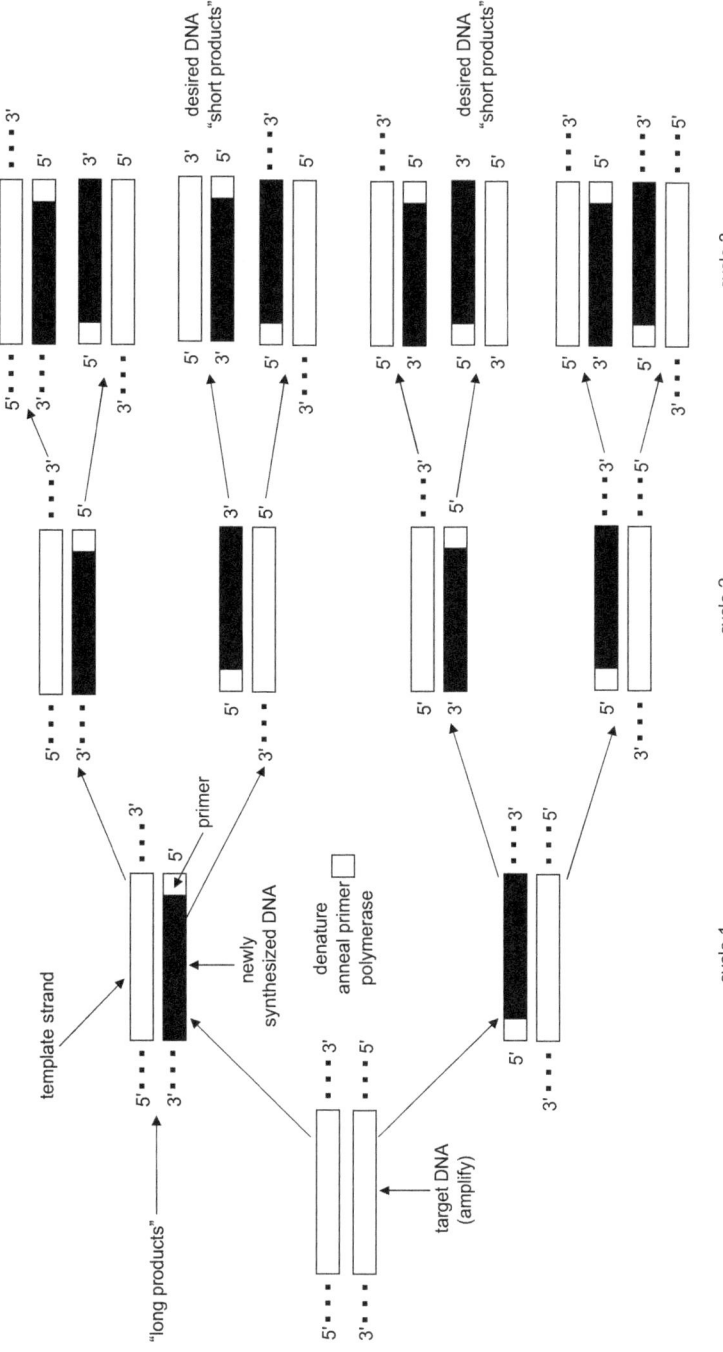

Figure 2.11 The first three cycles of PCR. The empty rectangles represent template DNA. The black rectangles indicate newly synthesized DNA. These will become template DNA in the next cycle. The three dots at the end of the DNA rectangles indicate DNA of varied length.

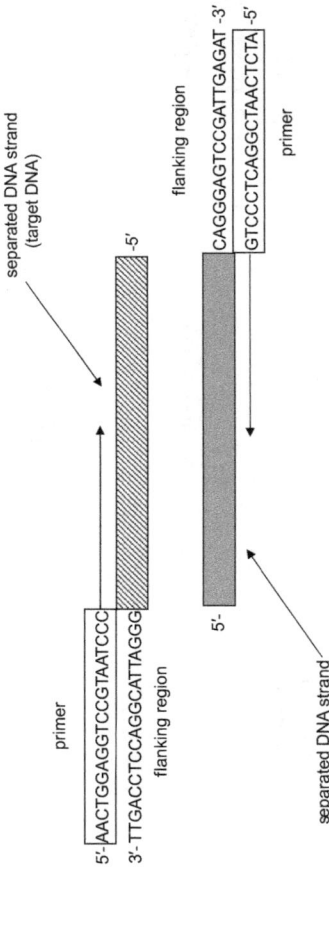

Figure 2.12 Primer design and target DNA. The primer is long enough for adequate specificity, so it will not hybridize to non-target sites but to the desired DNA fragment. It is short enough for quick template binding at the annealing temperature. The target DNA might very well be a short tandem repeat with flanking regions, which are usually conserved.

polymerase requires them in the $5'$ to $3'$ direction in order to produce a new strand.

There is a sequence of steps:

- **Denaturation.** The template DNA segment does not necessarily have to be a pure sample (it may be extracted from the cheek cells). It is heated to 94–98 °C for 30 seconds. The DNA strands separate.
- **Annealing.** The DNA is then cooled to 50–60 °C for 1 minute. This results in some rejoining of the single strand but, more importantly, permits the primers to attach to the target DNA.
- **Extension.** The temperature is raised to 72 °C for 2 minutes. This is a viable temperature for the operation of Taq DNA polymerase. New DNA strands are synthesized by the addition of the appropriate nucleotide (Figure 1.7; X = OH, Y = H) to the $3'$ end of the primers to give double-stranded DNA. The two long DNA strands cannot join because they are in very low concentrations.
- **Recycling.** The previous cycle is repeated many times (taking a few minutes each time) in an automated thermocycler. In the first cycle, a set of "long products" from each strand of the target DNA is obtained. These polynucleotides have identical $5'$ ends but random $3'$ ends (where the DNA synthesis terminates by chance). The second cycle results in the production of four double-stranded DNA strands, two of which are identical to the long products of cycle 1 and two of which are entirely new DNA. The third stage gives rise to authentic 100–1000 bp "short products", the $5'$ and $3'$ ends of which are both set by the primer annealing positions. The number of "short products" increases exponentially. Twenty cycles yield a million amplifications and thirty cycles yield a billion amplifications of the DNA. With each successive cycle, more and more of the desirable "short products", and relatively few of the undesirable "long products", are produced.

2.3.3 Emulsion PCR

This adaptation of the PCR process allows the sequencing of long lengths of DNA. The DNA is fragmented into 300–500 bp in length and separated into single strands. Each fragment is

Table 2.1 Comparison of vector cloning and PCR (rDNA, recombinant DNA).

Vector	PCR
More complex, needs rDNA made *in vitro* and amplified *in vivo*.	*In vitro* procedure and rDNA unnecessary.
Microgram quantities of highly purified DNA required for amplification.	Nanograms of DNA sufficient and DNA purity not critical.
Larger segments of DNA can be handled but errors are greater.	Can only deal with smaller lengths of DNA.
Can produce a desired protein at termination of the procedure.	Only amplified DNA obtained.
Relatively cheaper but requires days of tedious intensive labor.	Requires only hours and well suited to automation.

attached at each end to an adapter, which enables it (a) to be joined chemically to a very small coated metallic bead (20 μm long), as well as (b) providing the annealing sites for the PCR primers. PCR reagents and the DNA fragments attached to the beads are added to an oil–water mixture, which is emulsified. Millions of aqueous droplets are formed within the emulsion. It is arranged so that only one DNA template is attached to one bead and this occurs in each droplet so one fragment = one bead = one read. The DNA is amplified by PCR to give enough copies for sequencing. The emulsion is then broken. When the PCR process is completed the DNA is denatured and the beads carrying single-stranded DNA are transferred to the wells of a PicoTiter Plate. Each plate comprises 1.6 million wells and the diameter of each well is designed so that only a single capture bead will fit into the well. It is then usable in massively parallel sequencing (see Section 2.1.3). Both vector cloning and PCR are used extensively and each have their own advantages, which are outlined in Table 2.1.

CHAPTER 3

The Endogenous Manipulation of DNA Within the Cell

3.1 GENE EXPRESSION

Gene expression is at the very root of genetics involving, as it does, the transformation of information in a gene into a functional product. We are concerned in this chapter with structural genes (from Section 1.6.1 in Chapter 1), which yield proteins and represent the bulk of genes in humans. All proteins are encoded by genes but not all genes encode proteins. The latter genes express functional RNA products. Some genes (called house-keeping genes) are expressed (or operate) in all cells all the time. They are usually essential for cellular function. Some genes are expressed in specific cells. For example, genes that encode the vital muscle proteins actin and myosin are expressed only in muscle cells. This is achieved by regulating proteins (called transcription factors), which can switch the gene on and off. Expression of the correct gene in the right order and at the appropriate time is especially critical during embryonic development and differentiation (unspecialized cells or tissues in the embryo becoming specialized for specific functions). The conversion of DNA in a structural gene to protein is indirect and requires the mediation of RNA. It is surmised that this is necessary because the required single-stranded DNA would be more easily

Animal Genetics for Chemists
By Ralph G. Wilkins
© Ralph G. Wilkins 2017
Published by the Royal Society of Chemistry, www.rsc.org

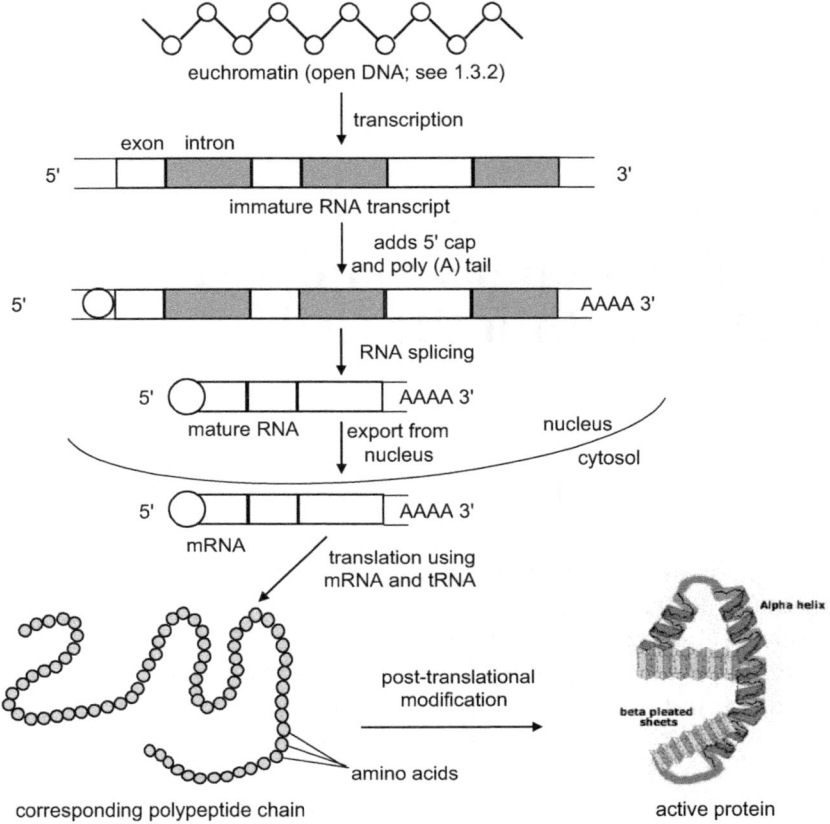

Figure 3.1 The steps in the conversion of DNA into a protein.

damaged than the DNA–RNA hybrid, which is quickly formed at the beginning of the overall process and is shown in Figure 3.1.

Some of the features of a structural gene that are used for the journey from the gene to one or more proteins are shown in Figure 3.2.

3.1.1 Types of RNA

A number of types of RNA, each generated by the corresponding gene, are involved at various stages in the production of protein from a gene. They include:

- Messenger RNA (mRNA), which is produced in the nucleus by transcription and used in translation to give

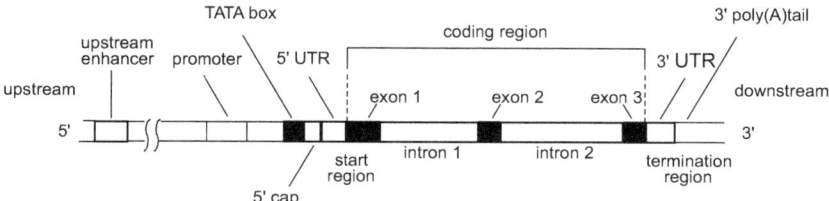

Figure 3.2 Anatomy of a structural gene. In general, introns are much longer than exons and will be removed before translation; exons represent the protein-encoding part of the gene. The untranslated regions (UTR) are non-coding but often contain regulatory elements that control protein synthesis. The upstream enhancer region may 10^3–10^6 bp away from the transcription start site but looped so as to be much nearer spatially to the promoter region. In addition to the features shown in the figure, there are also silencing regions, which are linked to repressor proteins, that slow or stop transcription. Insulator DNA segments protect a gene from the enhancer or silencer of a neighboring gene.

instructions for the sequence of amino acids in the polypeptide formed.

- Transfer RNA (tRNA), which is used for translation, where it helps in converting mRNA into the appropriate amino acid.
- Ribosomal RNA (rRNA), which is part of the ribosome at the site of translation.
- Small nuclear RNA (snRNA), which is used in post-transcriptional modification of pre-mRNA. In order to function, snRNA has to combine with a small number of proteins to form a small nuclear ribonucleoprotein (snRNP) complex. The functioning of these RNAs is aided by:
  - ○ RNA polymerases, which are used in all the stages of the conversion of a gene into a protein. There are three types, all complex enzymes, with several common features and high molecular weights. Type I in the nucleolus is involved in the synthesis of rRNA. Type II in the nucleoplasm aids in the synthesis of mRNA. Type III synthesizes tRNA and certain small RNA molecules.
  - ○ The ribosome, consisting of a small and a large subunit, which contain rRNA and a number of proteins. It functions at the translation site, using three binding sites in the large subunit. There are many ribosomes in a cell, either attached to the endoplasmic reticulum or free in

the cytoplasm (see Figure 1.2). Other RNAs described in Section 3.1.7 are important in the regulation of gene expression.

3.1.2   Transcription

Transcription is the first stage of gene expression in which a nucleotide sequence of a gene is transcribed within the cell nucleus into mRNA (Figure 3.3).

The process generally resembles the same process in prokaryotes but is more complex. Regions of the chromosome must be opened up so that transcription enzymes and factors can gain access to the gene. There are three major steps in transcription.

1. **Initiation:** RNA polymerase II recognizes and binds to a main binding site consisting of a TATAAAA sequence (TATA box; Figure 3.2), which is 19–27 bp upstream from the transcription site. This is near the first codon of exon 1.

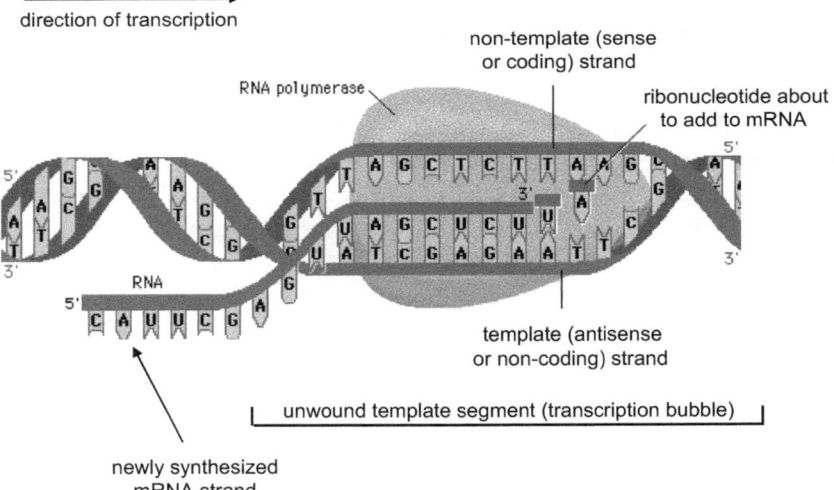

Figure 3.3   Transcription process. Addition of a complementary NTP (Figure 1.6, structure C) to the template DNA strand generates mRNA with release of a pyrophosphate ion. Uridine triphosphate U (UTP) and adenosine triphosphate A (ATP) are shown being added [compare with Figure 1.7 ($X = Y = OH$), which extends a DNA strand, rather than as here with an mRNA strand].

The start codon is AUG on RNA (ATG on DNA). RNA polymerase remains stationary during this stage and unwinds DNA a small degree to form two strands in the transcription bubble (Figure 3.3). Only the template (antisense) strand is affixed by the RNA polymerase and copied. It is the other non-template sense strand that contains the actual codons and may vary from one strand to the other along the DNA. Some genes are transcribed on one of the strands, some genes on the other strand and, sometimes, the same segment of DNA contains genetic information on both strands.

2. **Elongation:** RNA polymerase moves along the template strand and adds a ribonucleotide triphosphate (NTP) to the messenger RNA that is complementary to the nucleotide on the template strand (Figure 3.3). As it moves, the RNA polymerase shuts the transcription bubble containing the transcribed DNA (now rewound) and opens a new unwound DNA region. Knotting is prevented by topoisomerases. Nucleotides (from a nucleotide pool) complementary to the template DNA will add to the mRNA, which will be continually synthesized. The entire gene, containing both introns and exons, is transcribed until termination, when RNA polymerase reaches the end of the gene.

3. **Termination:** In most mammalian protein-coding genes that we are considering here, RNA polymerase can terminate at multiple sites 0.5–2 kb beyond the poly(A) sites (Figure 3.2). The exact relationship between RNA polymerase II, mRNA release and termination is uncertain.

3.1.3 Post-transcriptional Processing: Splicing

The so-called pre-mRNA, resulting at termination, contains a number of fragments (introns), which are not used in the subsequent treatment (translation) but have been shown to affect gene expression. Exons are usually found at the ends of the gene and the ends usually contain sequences that are not translated (5′ UTR and 3′ UTR). The non-coding introns must be removed and the remaining exons strung together (splicing) to give the mature mRNA, which exits the nucleus and is translated into polypeptides and subsequently proteins in the cytoplasm. This is aided by spliceosomes.

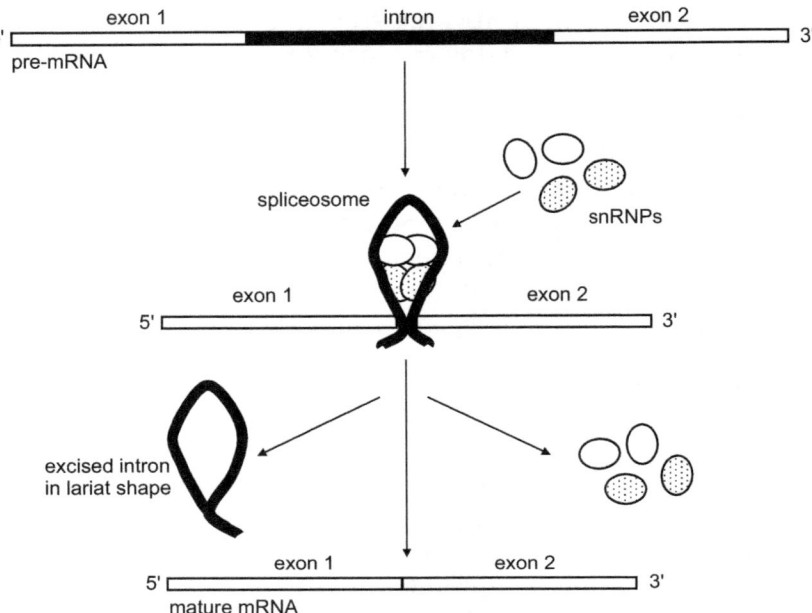

Figure 3.4 Excision of an intron from pre-mRNA to give mature mRNA.

A spliceosome is a large complex of different small nuclear ribonucleoproteins (snRNPs), pronounced "snurps"! The snRNPs have different roles within the spliceosome, which recognizes sequences at the beginning (GU) and end (AG) of the intron, attaches to these, forms a loop, snips out the intron and finally ties the two encompassing exons together—an impressive performance (Figure 3.4). The resulting mature mRNA without the introns will, of necessity, be much smaller than the pre-mRNA, for example, about 25% in the case of the insulin gene and less than 1% (2.5 mb pre-mRNA reduced to 14 kb mature mRNA) with the dystrophin gene.

3.1.4 Alternative Splicing

In humans, a number of the genes undergo splicing in which all the exons are retained and all the introns removed. However, many genes undergo alternative splicing, in which all the introns are removed, along with some of the exons.

- The *dystrophin* (*DMD*) gene on chromosome Xp is the largest known human gene. There are 79 exons in its pre-mRNA and 78 large-size introns. Slightly different alternative splicing of the pre-mRNA leads to numerous functionally related dystrophin proteins, differing slightly in their amino acid sequences (isoforms). Deletion of one or more exons may lead to Duchenne and Becker muscular dystrophies (see Section 6.3.5 in Chapter 6).
- The primary transcript for the *calcitonin* gene contains six exons (Figure 3.5).

There are two modes of splicing used to produce the mature mRNA. In both, all the introns are removed. In one case, exons 5 and 6 are removed, and poly(A) added at the end of exon 4; and, in the other case, only exon 4 is removed and poly(A) attached at the end of exon 6. This leads, on translation, to calcitonin with 32 amino acids and calcium-gene-related peptide (CGRP) with 37 amino acids (Figure 3.5). There are similarities but also significant differences between the two products. Calcitonin is prevalent in the thyroid gland and has a significant role in calcium and phosphorus metabolism. CGRP is produced in neurons and

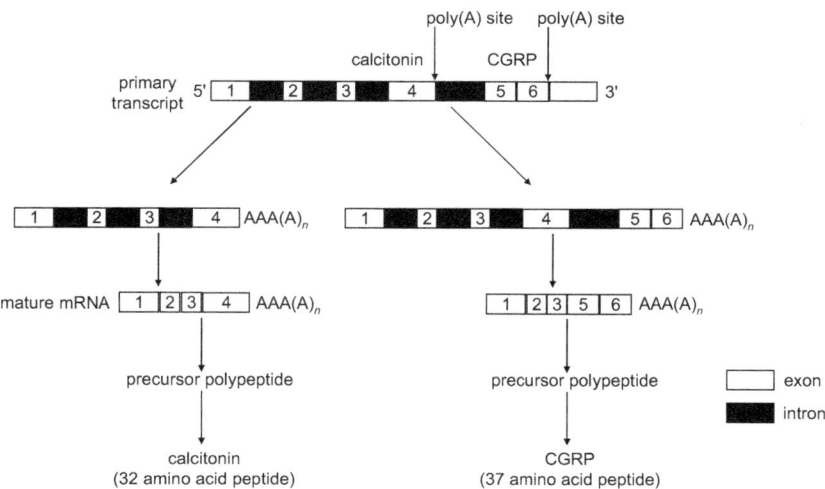

Figure 3.5 Alternative pre-mRNA splicing showing the complete path from the primary transcript to the protein products for the calcitonin gene. Introns are represented by black rectangles.

vascular systems, and plays a key role in triggering migraine by widening blood vessels near the neurons in the head. Encouraging results in moderating the frequency and intensity of migraines have been obtained by injecting monoclonal antibodies specifically engineered to wipe out the offending CGRP. Other alternative splicing modes are encountered. Some of the introns, as well as the exons, may be retained. Incorrect splicing positions may be used. All modes of alternative splicing lead to a multiplicity of arrangement of the genes in the mature mRNA, and therefore a variety of polypeptide products or indeed even no product at all. Alternative splicing produces, on average, about 10 different proteins from each gene. Alternative splicing is a natural process and it is estimated that as many as 80% of human genes undergo alternative splicing. Obviously, cutting and splicing by the spliceosome must be very precise, otherwise nucleotide misalignments occur and mutations result. Many genetic human disorders arise from alternative splicing and abnormal splicing. One example is spinal muscular atrophy.

Spinal Muscular Atrophy

The *survival of motor neuron 1, telomeric* (*SMN1*) gene on human chromosome 5q encodes the survival motor neuron (SMN) protein. The protein is important in processing pre-mRNA and in preventing incorrect splicing. It helps maintain effective motor neurons in the spinal cord, which control muscle movement. Mutations may delete exon 7 in both copies of the *SMN1* gene in each cell as a result of inefficient splicing. Little or no SMN protein results. Ninety-five percent of individuals with spinal muscular atrophy have this mutation. All muscles become weaker and death before the age of two years occurs in 50% of children with this condition. It is the leading genetic cause of infant mortality. An *SMN2* gene located close to *SMN1* encodes the same protein but in much smaller amounts and its RNA transcript is also prone to incorrect splicing. Orally active drugs are currently being sought to target spliceosomes linked to splicing malfunctions, which may lead to spinal muscular atrophy, as well as cancers and eye problems, which account for some 10% of genetic diseases.

3.1.5 Translation

For translation, the final mature mRNA needs the $5'$ cap for recognition in this step and a poly(A) tail to protect the mRNA from enzymatic (nuclease) degradation (Figure 3.2). In the cytoplasm of the cell, mature mRNA (hereafter simply called mRNA) is translated into a sequence of amino acids to form a polypeptide chain. This usually needs some processing to give a final functional protein. Three consecutive nucleotides, termed a codon triplet or codon, in the mRNA equate to an amino acid (Figures 3.6 and 3.7). Since the mRNA is derived from DNA during transcription, the code can also be written, and this is more usual, in terms of the DNA from which the mRNA is derived.

Translation takes place on a ribosome in the cytoplasm after the poly(A)–mRNA moves out of the nucleus into the cytoplasm. Small (40S) and large (60S) subunits of the ribosome assist in the translation. The large subunit has three binding sites, designated A (acceptor), P (peptidyl) and E (exit), which are used during translation, Figure 3.8.

Transfer RNA (tRNA; Figure 1.8) is the important vehicle responsible for converting the mRNA codon into the appropriate amino acid. It does this by covalently linking an amino acid to the $3'$ end of the tRNA to give an aminoacyl-tRNA (Figure 3.8A). There are 20 different aminoacyl-tRNAs, each carrying a different amino acid. The anticodon is a triplet of nucleotides in the tRNA, which bind by hybridization to a specific triplet of nucleotides (a codon) on the mRNA. The stages in the process are as follows:

- **Initiation:** The smaller ribosome subunit is loaded with several initiating factors, as well as an aminoacyl-tRNA carrying the amino acid methionine. When it encounters the mRNA, the smaller ribosome unit binds to a site upstream (on the $5'$ side) of the triplet AUG, which is the starting point, and moves downstream until it locates it. Now the larger ribosome subunit is joined to the smaller and translation is initiated. This initial methionine-loaded tRNA occupies the P site on the complete ribosome (Figure 3.8A). The region between the RNA cap and the

RNA

first base	second base U	C	A	G	third base
U	UUU Phe	UCU Ser	UAU Tyr	UGU Cys	U
	UUC Phe	UCC Ser	UAC Tyr	UGC Cys	C
	UUA Leu	UCA Ser	UAA Stop	UGA Stop	A
	UUG Leu	UCG Ser	UAG Stop	UGG Trp	G
C	CUU Leu	CCU Pro	CAU His	CGU Arg	U
	CUC Leu	CCC Pro	CAC His	CGC Arg	C
	CUA Leu	CCA Pro	CAA Gln	CGA Arg	A
	CUG Leu	CCG Pro	CAG Gln	CGG Arg	G
A	AUU Ile	ACU Thr	AAU Asn	AGU Ser	U
	AUC Ile	ACC Thr	AAC Asn	AGC Ser	C
	AUA Ile	ACA Thr	AAA Lys	AGA Arg	A
	AUG Met	ACG Thr	AAG Lys	AGG Arg	G
G	GUU Val	GCU Ala	GAU Asp	GGU Gly	U
	GUC Val	GCC Ala	GAC Asp	GGC Gly	C
	GUA Val	GCA Ala	GAA Glu	GGA Gly	A
	GUG Val	GCG Ala	GAG Glu	GGG Gly	G

DNA

first base	second base T	C	A	G	third base
T	TTT Phe	TCT Ser	TAT Tyr	TGT Cys	T
	TTC Phe	TCC Ser	TAC Tyr	TGC Cys	C
	TTA Leu	TCA Ser	TAA Stop	TGA Stop	A
	TTG Leu	TCG Ser	TAG Stop	TGG Trp	G
C	CTT Leu	CCT Pro	CAT His	CGT Arg	T
	CTC Leu	CCC Pro	CAC His	CGC Arg	C
	CTA Leu	CCA Pro	CAA Gln	CGA Arg	A
	CTG Leu	CCG Pro	CAG Gln	CGG Arg	G
A	ATT Ile	ACT Thr	AAT Asn	AGT Ser	T
	ATC Ile	ACC Thr	AAC Asn	AGC Ser	C
	ATA Ile	ACA Thr	AAA Lys	AGA Arg	A
	ATG Met	ACG Thr	AAG Lys	AGG Arg	G
G	GTT Val	GCT Ala	GAT Asp	GGT Gly	T
	GTC Val	GCC Ala	GAC Asp	GGC Gly	C
	GTA Val	GCA Ala	GAA Glu	GGA Gly	A
	GTG Val	GCG Ala	GAG Glu	GGG Gly	G

Figure 3.6 The genetic code for nuclear RNA and DNA. This shows the amino acids corresponding to the 64 combinations (codons) possible using the four nucleotides that comprise RNA and an equivalent chart showing the DNA codons and their corresponding amino acids. The code differs slightly for mitochondrial DNA (see Section 3.1.10). Some ribonucleotide triplets do not express an amino acid but a stop codon instead, signaling the termination of translation. Most amino acids have more than one triplet combination.

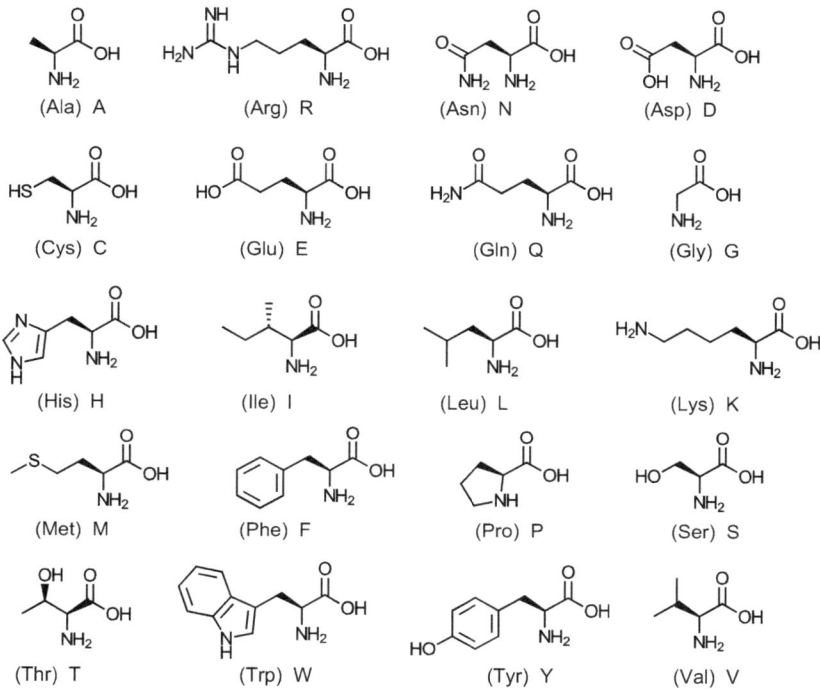

Figure 3.7 Names, abbreviations and structures of the 20 biologically important amino acids.

codon AUG is termed the 5′-untranslated region (5′UTR; see Figure 3.2). Once translation has started, many ribosomes can start at different points along the mRNA, similar to the many origins of replication in DNA replication (see Section 3.3.1).

- **Elongation:** Assisted by elongation factors, we are now ready to convert the next codon adjacent to AUG on the mRNA (GUC; Figure 3.8B) into an amino acid. Only an aminoacyl-tRNA that has an anticodon for GUC, namely CAG, and bearing the amino acid valine (Figure 3.6) will bind to the mRNA and this is at the A site of the ribosome. At the same time, the methionine-tRNA bond is broken, and the methionine and the valine are joined to form a peptide bond, which is catalyzed by a peptidyl transferase enzyme in the ribosome. The now "empty" tRNA (deacylated, *i.e.*, without

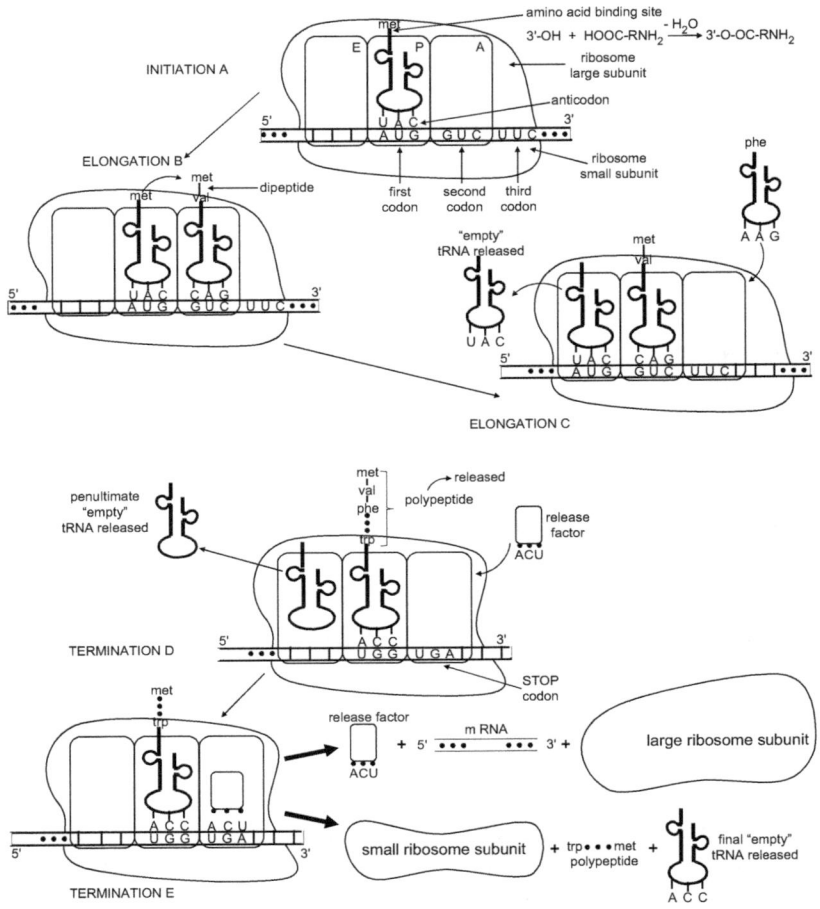

Figure 3.8 Important entities (A–E) in the three stages of translation.

the amino acid) will leave the ribosome at the E site (Figure 3.8C). The ribosome moves forward again and attaches to the next codon, which is UUC, and a new aminoacyl-tRNA with the anticodon AAG and bearing phenylalanine will attach to the mRNA at the A site and the process is repeated, adding phenylalanine to the growing chain. This continues for many codons, building up a polypeptide chain until (as it turns out) the final amino acid (Trp) is added (Figure 3.8D).

- **Termination:** When the traveling ribosome reaches the codon UGA, which is a stop codon (UAG and UAA are others; Figure 3.6), translation ceases (Figure 3.8D). A release factor with the anticodon ACU now binds to the ribosome and this stimulates the release of the polypeptide from the P site, as well as to ease the removal of all the components of the ribosome from the mRNA (Figure 3.8E). The released ribosome splits into its subunits for reuse in another round of translation. The methionine at the start of the polypeptide chain is usually removed. The segment between AUG and the stop codons is termed an open reading frame (ORF). Nucleotides from the stop point to the poly(A) tail make up an untranslated region (3'UTR; see Figure 3.2).

3.1.6   Post-translational Processing

The polypeptide at the end of translation needs to go through a number of steps to give the final functional protein. Usually, the native conformation of the protein is produced, aided by chaperone proteins. Misfolded protein can be destroyed by the cell, but this is more difficult as a person ages and this often heralds protein aggregation, which is possibly the basis for neurodegenerative diseases such as Alzheimer's and Parkinson's. The protein is moved to the cytoplasm if it is water soluble, or to the cell membrane if it is hydrophobic, possibly for further processing. This could entail a variety of treatments. Some proteins, for example, are only active if they are phosphorylated on particular amino acids. This represents a means for regulating protein production. Single polypeptide chains from the translation process may be rearranged and converted into functional proteins, which is well illustrated by a consideration of insulin.

Human Insulin

The *INS* gene on human chromosome 11p is one of many genes associated with type 1 diabetes. The *INS* gene provides instructions for producing the human hormone insulin, which is essential for controlling the level of blood sugar. The primary translation product of the gene is preproinsulin (Figure 3.9), which has 110 amino acids. It is relatively inactive. Amino acids

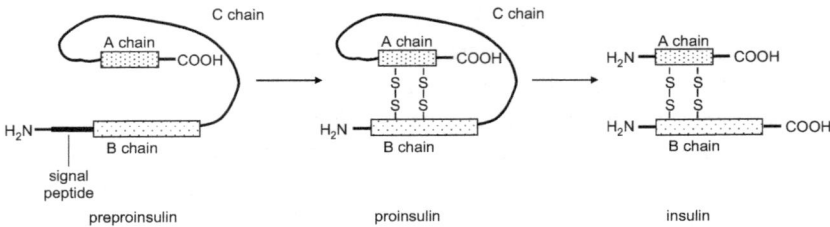

Figure 3.9 Post-translational treatment of preproinsulin in the final generation of insulin.

are removed by enzyme digestion to give proinsulin. This folds and two disulfide bonds between the cysteines are formed. Protease cleaves proinsulin with further loss of amino acids to give mature and active insulin, now containing only 51 amino acids, one of the smallest proteins in the human body.

3.1.7 Proteins

Proteins are involved in many of the processes carried out in cells. This includes gene expression and structural, functional and regulatory control of the body tissues and organs, and eventually the physical characteristics of the animal. There are many types of proteins, which can be classified in a variety of ways, most of which occur in the book (see Table 3.1).

3.1.8 The Regulation of Gene Expression

It is, of course, necessary for gene expression to be regulated so that the amount or nature of protein resulting can be controlled. There are a number of places during the overall gene-to-protein conversion where this can take place. Some of these have been pointed out. Alternative splicing (Section 3.1.4) leads to different mRNAs and therefore different (often many) proteins, if required. Epigenetic modifications of DNA (Section 7.4.2 in Chapter 7) allow access to the genes encoding proteins involved in activation or repression of the gene. We have seen that a number of RNA molecules play important roles in the implementation of transcription and translation. Others are involved in the regulation of gene expression.

Table 3.1 Types of protein, with descriptions and examples.

Type	Description	Example
Contractile	Mediate contractile processes in cytoskeleton, and cardiac and skeletal muscles.	Actin and myosin in mitosis and meiosis spindle function. Form the contractile ring in cytokinesis (see Section 3.4 and Section 5.1 in Chapter 5).
Enzyme	Catalyze most of the classical reactions in the body. Assists in the production of proteins.	Three enzymes involved in the metabolism of phenylalanine (see Section 6.2.11 and Figure 6.11 in Chapter 6).
Hormone	Proteins, usually globular, water-soluble and quick acting, unlike steroids. Chemical messengers coordinating body processes.	Pancreatic insulin (Section 3.1.6) helps regulate blood-sugar levels.
Membrane	Different types connected with the cell plasma membrane. Allows passage of chemicals across membrane, linking intra- and extra-cellular structures.	Transmembrane CFTR protein forms a channel in the membrane for chloride ion transport (see Section 4.3.7 in Chapter 4).
Storage	Maintain reserve of metal ions and amino acids necessary for maintenance and growth of the animal.	Ferritin, the iron storage protein. Iron imbalance results in hemochromatosis (Section 6.2.11 in Chapter 6).
Structural	Provide support for cell walls, skin, bones, hooves, *etc.*	Collagen. Impaired collagen production results in osteogenesis imperfecta (Section 7.2.5 in Chapter 7).
Transport	Bind and carry molecules to every part of the body.	Hemoglobin in blood binds oxygen for use in the body. A variety of diseases are associated with mutations, *e.g.*, sickle-cell anemia (Section 6.2.11 in Chapter 6).

RNA Interference (RNAi)

This is an important natural process for regulating gene expression and finds significant use for directed gene silencing (knockdown) in the laboratory. Using it, a gene function can be put on temporary hold and thereby examined at leisure.

- The RNAi process is initiated by a long, double-stranded RNA (dsRNA) originating in the nucleus. This binds to the enzyme Dicer in the cytoplasm and is cleaved into smaller fragments of short (or small) interfering RNA (siRNA), which are 20–25 nucleotides long with dinucleotide overhangs.

- This siRNA duplex is then unwound and separated into a single guide strand (antisense; Figure 3.3) and a passenger (degraded) strand by associating with a helicase- and endonuclease-containing protein complex (RNA-induced silencing complex, RISC).

- The RISC with the guide strand can then bind to complementary sequences in a *small* part of the mRNA of a target gene. This results in endonucleolytic cleavage of the mRNA at a specific site in the middle of the duplex and thereby mediates its silencing. It effects post-transcriptional regulation of gene expression and, as a consequence, functional RNA or protein is translated at lower levels (it is knocked down). The genetic code is not changed.

- In an exogenous application, siRNA is synthesized and designed to specifically target (by hybridization) an mRNA that it is desired to silence. The synthetic double-stranded siRNA is introduced into cells that are amenable to transfection, for example, by electroporation. Then, the RNAi machinery is used to silence the gene. The function of most of the genes in the small roundworm *Caenorhabditis elegans* (the first animal used to characterize the RNAi mechanism) has been examined using this approach. The results give some indication of how fat is stored and obesity is controlled in the animal, properties relevant to humans since about 70% of the proteins used are common to the two. Although the technique has been highly important in basic research, it has until recently been less useful for gene therapy (Section 8.4.1 in Chapter 8). This is changing and RNAi-based approaches for liver-linked diseases, in particular, are being actively pursued.

As well as siRNA and dsRNA, other key substrates used for gene regulation are miRNA and asRNA.

- MicroRNA (miRNA) is similar in length to siRNA and helps regulate and transcribe genes. Double-stranded miRNA activates the RNAi machinery to block protein synthesis by suppressing translation.

• Antisense RNA (asRNA) is single-stranded RNA with a nucleotide sequence complementary to a sense mRNA strand. It is naturally occurring or may be synthesized. Combined with sense RNA, antisense RNA may silence gene expression at the transcription, post-transcription or translation stages.

3.1.9   Environmental Effects on Gene Expression

Most genes, and therefore traits, are minimally affected by the external environment and much more by internal influences. Diet and nutrition (Sections 7.4.1 and 7.4.4 in Chapter 7) and exogenous toxic chemicals (Section 4.6 in Chapter 4) are factors that may, however, determine a gene expression and animal phenotype. An excellent and rare example of a temperature influence on gene expression is afforded by the Himalayan rabbit (see Siamese cats in Section 7.3.9). Environmental factors, in addition to genetic factors, do in fact determine phenotypes (disorders, *etc.*). Why else would identical twins with identical gene contents finish up with different diseases?

3.1.10   Mitochondrial DNA

Transcription and translation of mtDNA resembles that in bacteria. One or other of the two strands of mtDNA are transcribed along the entire circle of mtDNA. The majority of the mitochondrial genes are transcribed on the heavy (H) strand. Transcription starts at the H-strand promoter (HSP) in the control region (Figure 1.14 in Chapter 1) and proceeds counterclockwise. It transverses two rRNAs, twelve mRNAs and fourteen tRNAs (Section 1.7.4 in Chapter 1). Light (L) strand transcription starts at the L-strand promoter (LSP) near the HSP in the control region and proceeds clockwise, coding for one mRNA and eight tRNAs (Section 1.7.4 in Chapter 1). The long polycistronic (encoding more than one protein) RNA chains formed are subsequently cleaved into separate RNA entities. Many proteins required by the mitochondria are synthesized by nuclear genes and then transported to the mitochondria.

The genetic code for mtDNA is slightly modified from that for nDNA. For example, the usual start codon with nDNA transcription (AUG) is sometimes replaced by AUA or AUU for mtDNA. The codons AGA and AGG (and not UGA) are used as stop codons in mtDNA.

3.2 THE CELL CYCLE

The cell cycle is a precise sequence of events. It leads to two cells, each identical to the one starting cell (including the DNA content). It is vital because cells can only be synthesized from existing ones. The trillion (10^{12}) cells making up an animal originate from a single fertilized egg cell. Animal survival requires that millions of cells, for example, about two million red blood cells in a healthy person, are produced every second by cell cycles and, in the process, replace dying cells. Defects in the cell cycle, particularly in its control, may lead to chromosomal defects and animal disorders.

There are five main phases and these are termed G1, S, G2 (collectively termed the interphase), M and C (Figure 3.10). Each phase must be completed before the next phase can begin. In the interphase, DNA is a tangle of light-staining (euchromatin) and dark-staining (heterochromatin) fibers occupying most of the nucleus (Section 1.4.3 in Chapter 1). In the dividing cell (mitosis), the chromatin is more compacted to the classical chromosome shape (Figure 1.10).

- **Growth (G1) phase:** Within this phase there is a high rate of biosynthesis and cell growth. When the cell reaches a certain size, it enters the S phase.
- **DNA synthesis (S) phase:** In this phase the DNA content of the cell doubles and the chromosomes duplicate (Section 3.3.1).
- **Growth (G2) phase:** By the late stages of G2 and the early part of the next stage (M), chromatin has condensed to a definite shape. This is another period of growth and final preparations are made for cell division. During the interphase, the cell carries out most of its normal functions.
- **Mitosis (M) phase:** Both nuclear division (mitosis; see Section 3.4) and the beginnings of cytoplasm division

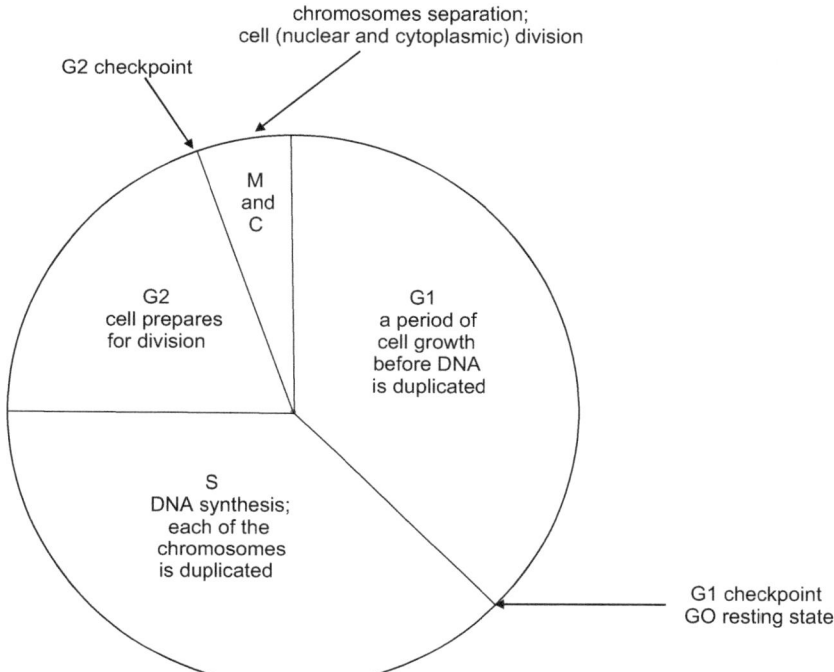

Figure 3.10 The main phases in the cell cycle. In human epithelial-like HeLa cells, the phases take 8.2, 6.1, 4.6 and (M plus C) 0.6 hours, respectively.

(cytokinesis) are involved. In this stage the chromatin is microscopically visible as the X-shaped chromosome. In mitosis, the two sister chromatids in each chromosome are separated and segregated to end up in the two "daughter" cells. This means that the chromosome set up in each of the daughter cells is identical to that in the original cell before duplication.

- **Cytokinesis (C) phase:** This phase is often included with mitosis, although the two can be considered as separate stages. Two separate cells commence to form during the late stages of mitosis and the plasma membrane in the middle of the cell is drawn in to form a cleavage furrow, which finally becomes two complete cells. Mitosis and cytokinesis, and the onset of the S-phase, are key steps in cell division.

The pattern of events is basically the same with all animals, although the cycle time will vary. The usual cell cycle is 10–30 hours, with more than 90% of that time spent in the interphase. The duration of the cycle may also vary with different tissue types, *e.g.*, in humans, adult nerve and muscle cells do not divide at all. Liver cells divide once a year.

3.2.1 Exit and Checkpoints

It would be surprising if there were not places along the cell cycle where its progress could be assessed or indeed terminated (Figure 3.10). Towards the end of the G1 phase, a checkpoint assesses whether the cell has grown sufficiently, has synthesized the appropriate proteins and whether there has been any damage to its DNA. If there are problems, the cell may "rest" in the GO state (Figure 3.10) until ready to re-enter or be permanently retired. A second checkpoint during interphase is made at the end of the G2 phase. Again, cell size and any DNA damage is checked, as well as being verified that replication of DNA has occurred in the S phase. Only when this checkpoint is passed does the cell initiate mitosis. In mitosis, a final check is made to determine if the correct attachment of the chromosome to the spindle assembly has been made. If checkpoints are ignored or overridden, uncontrolled cellular growth may result, *i.e.*, the possible development of a cancer cell. The molecular mechanisms that regulate these controls are quite complex. Double-strand break repair (Section 4.6 in Chapter 4) is invoked during the cell cycle.

3.2.2 Cell Division and Death

Cells grown in cultures replicate a number of times before mitosis is terminated and the cell dies. There are four stages:

1. First, there is a rapid, healthy division of the cell.
2. Then, a slowing down of mitosis occurs.
3. The cells stop dividing in a process termed senescence.
4. Shortly after, the cells commit suicide by a triggered programmed cell death (apoptosis). The apoptosis pathway is tightly regulated so that cell death occurs only under

Table 3.2 Average age at death and number of senescence cell divisions for various animals.

Animal	Average lifetime (years)	Average number of cell divisions before senescence
Mouse (lab)	2	20 ± 6
Chicken (newborn)	12	25
Chicken	30	25 ± 10
Human	76	50 ± 10
Human with progeria	13	6 ± 4
Galapagos turtle	175	115 ± 9

specific circumstances. It is a very complex process, involving a number of proteins. How many times does an animal cell divide before it gives up? In humans, the average number of cell doublings, *via* cell division, before the cells stop dividing and finally commit suicide is 40–50.

There does appear to be a correlation between the average age of death of an animal and the average number of cell divisions before senescence sets in for that animal. This is often called the Hayflick limit. It applies to a number of animals besides those shown in Table 3.2. The concept is well illustrated by a comparison of humans without or with the very rare autosomal dominant Hutchinson–Gilford progeria syndrome (Section 6.2.1 in Chapter 6).

The *lamin A/C* (*LMNA*) gene on chromosome 1q encodes the protein lamina, which is important in shaping the cell nucleus. A specific C1824T mutation leads to a shortened (by 50 amino acids) protein and the resultant progeria syndrome involves rapid aging from childhood.

3.3 DNA REPLICATION IN THE INTERPHASE

3.3.1 Nuclear DNA Replication

DNA is replicated (duplicated) during the interphase of each cell cycle and the process is absolutely essential and complex, and involves a set of proteins, including enzymes. DNA replication is generally similar for nDNA and mtDNA, despite their having different linear and circular structures, respectively. A simple representation for nDNA is shown in Figure 3.11. Only a portion

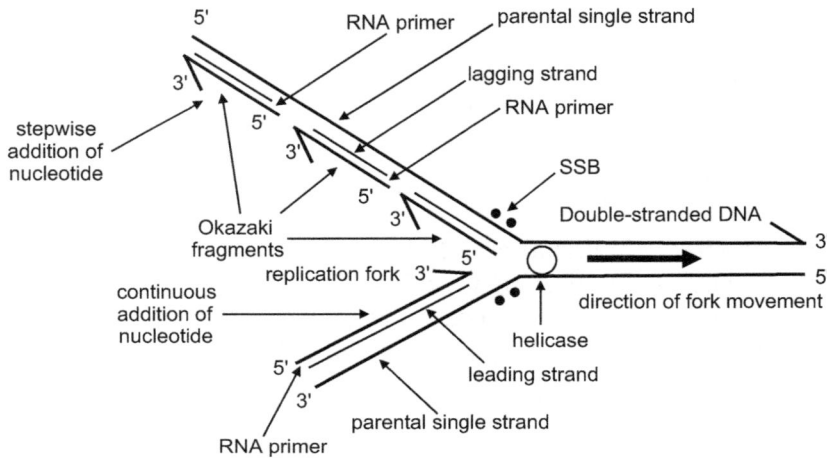

Figure 3.11 A simplified representation of the replication of DNA.

of the target DNA is replicated at a given time so that the two DNA strands are never completely separated, when they would tend to be more unstable. The process can be considered as three steps.

1. **Initiation:** Replication is initiated at a particular sequence—the origin of replication. The two strands of the DNA helix are separated into two single strands using helicase to break the hydrogen bonds between the base pairs. This is easier in places that are rich in A–T pairs, which only have two hydrogen bonds (see Section 1.2.2 in Chapter 1). The single strands are stabilized by single-strand DNA-binding proteins, which define the replication fork (Figure 3.11). Topoisomerases prevent the tangling of the free single DNA strands. Before the strand duplication can be started in earnest, it is necessary to attach a short segment of a nucleotide with a free terminal –OH group to the initial part of the DNA segment that is to be copied. This is an RNA primer consisting of a 5–15 bp ribonucleotide complementary to the DNA segment and synthesized by

primase formed in the cell. Deoxynucleotides can now be added by DNA polymerase to generate the new strand, as outlined in Section 1.2.4 in Chapter 1.

2. **Strand synthesis:** Each parental single strand acts as a template for the production of a new strand that is complementary to the original strand. As each nucleotide on the strand is exposed, a complementary nucleotide is added from a pool of dNTPs. The reaction is catalyzed by DNA polymerase III. The incorporation reaction is shown in Figure 1.7 ($X = OH$, $Y = H$). DNA polymerase can only effect the addition of new nucleotides to the $3'$ end of an existing strand, and therefore the synthesis of new DNA must be in the $5'$ to $3'$ direction. This is no problem in the formation of one of the strands (leading) and nucleotides can be inserted in a smooth, continuous manner ($5'$ to $3'$) to generate a new strand using a single RNA primer. This is not the case in the formation of the other, separated template strand (lagging). It is running in the opposite, wrong ($3'$ to $5'$) direction. A more complex, discontinuous process must be used to synthesize this strand. A series of short segments of new DNA (100–200 bases long), each with its own RNA primer, is generated in the $3'$ to $5'$ direction at several places along the unwound strand using DNA polymerase III. The DNA fragments (Okazaki fragments) are finally linked using the enzyme ligase. Each primer is replaced by a complementary segment of DNA using DNA polymerase I at the appropriate time. As the replication fork continues, the leading and lagging strands, which have been opened, are twisted back into the DNA helical form.

3. **Termination:** The array of replication proteins and enzymes used exist permanently in the nucleus, but little is known about the details of the termination process. It occurs at specific sites where there is a unique nucleotide sequence. Two replication forks are formed at the origin of replication and are extended in both directions to give a replication bubble that enlarges and eventually fuses with other replication bubbles to give a continuous duplicate DNA. This is possible because very many origins of replication are initiated mostly at the same time, as many as 20 000 in human DNA duplication. The multi-positional replication is

essential. The copying of 80 million base pairs in an average human chromosome at the rate of about 50 base pairs per second, using just one origin of replication, would take about a month. The observed time is about 1 hour, as a result of the very many sites where replication can begin.

Replication Compared with Transcription

The replication process resembles transcription (Section 3.1.2) but with several important differences:

- Three basic steps (initiation, elongation and termination) are used in both.
- Both use single-stranded DNA as a template, although two strands are processed with replication.
- Both add a succession of nucleotides to a single DNA strand but use dNTP with replication and NTP with transcription. Thus, dsDNA is formed in replication using a DNA polymerase and ssRNA is formed in transcription using an RNA polymerase.
- Replication requires a primer to initiate the process, but this is unnecessary in transcription.
- Replication uses A–T and G–C base pairing, whereas transcription uses A–T (or A–U) and G–C base pairing.
- The entire genome is copied in replication, but transcription only copies a portion of a genome.

There are also many similarities with the PCR process described earlier (Section 2.3.1 in Chapter 2). DNA replication is, however, a natural process producing full-length DNA, whereas the PCR process is confined to the laboratory and a segment of DNA is replicated defined by primers.

3.3.2   Mitochondrial DNA Replication

It is perhaps surprising that the mechanisms of replication (and transcription; see Section 3.1.10) should be poorly understood for such a relatively simple molecule.

In the same way as nDNA replication, separate DNA strands are required and two new strands must be synthesized by complementary base pairing. In the long-standing strand displacement model, leading and lagging strand syntheses are uncoupled (compared to tandem replication in nDNA). RNA primer initiates replication of the leading strand at the origin of replication (oriH; Figure 1.14) after helicase opens the double-stranded molecule. When leading strand synthesis has proceeded about two-thirds of the circle, it exposes the second site (oriL) and lagging strand synthesis proceeds in the opposite direction. Both strands are completely transcribed along the entire genome. Replication proceeds without resorting to the use of Okazaki fragments. Current thinking is that the strand displacement model should be replaced, or at least augmented, by other models involving coupled leading and lagging strand synthesis, as with bacteria.

3.3.3   Errors and Repair in DNA Replication

DNA replication is amazingly accurate, it being estimated that only one uncorrected nucleotide error occurs during each cell division. There are a number of places where errors might occur. Sometimes, polymerases insert too many, or too few, nucleotides because a small loop develops during replication (strand slippage; Figure 3.12).

In the illustration in Figure 3.12A, mispairing only involves one CAG repeat. Several or even many repeats could be involved. If this occurs in the coding region it could lead to a polyglutamine disease. This type of error may arise, for example, when the normal 10–35 repeats of a CAG triplet is inadvertently expanded by strand slippage to 36–120 repeats, leading to Huntington disease (see also other trinucleotide expansion disorders in Section 4.4.1 in Chapter 4).

Replication Mismatch Repair

During replication, mismatching of the two nucleotides may occur. A guanine base on the template strand may pair with a thymine base on the newly synthesized daughter strand, rather than a required cytosine base. Repair must be made in the daughter strand where the error has occurred. This is

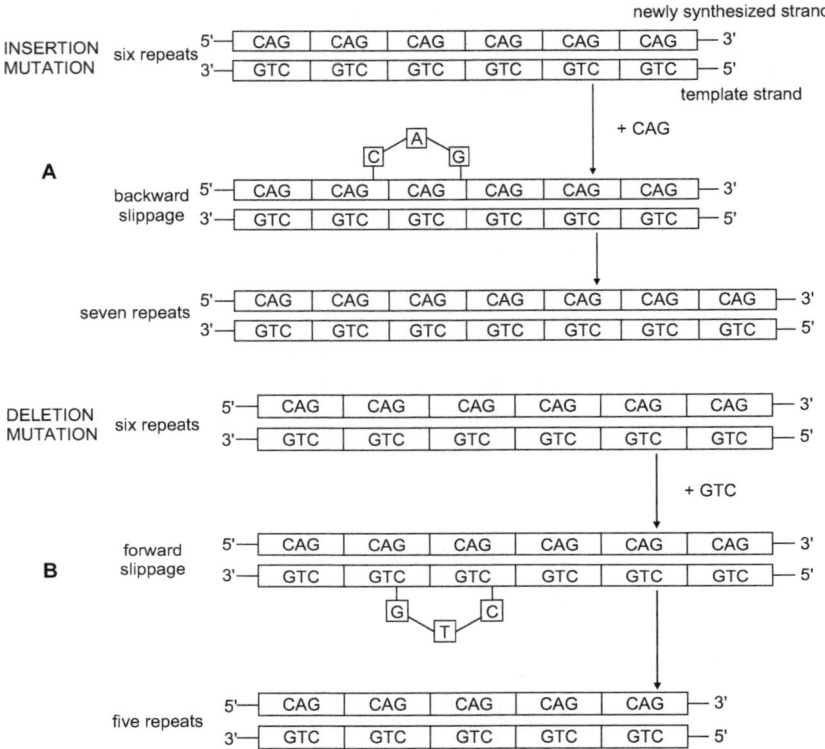

Figure 3.12   (A) A newly synthesized (replicating) strand slips or wrinkles (backward slippage) so that extra nucleotides are inserted (insertion mutation). (B) A wrinkle forms in the template strand (forward slippage) and nucleotides are omitted from the replicating strand (deletion mutation).

accomplished by a number of steps with the aid of various protein complexes and enzymes.

- There is immediate recognition of the location of the mispaired bases in the replication fork (Figure 3.11).
- An endonuclease cleaves the daughter strand containing the error. In *E. coli* only the parent strand is methylated (Section 7.4 in Chapter 7), so the cutting enzyme attaches to the unmethylated strand and cuts it. How the error-containing strand in human mismatch repair is identified is uncertain, but methylation signposts may not be used.

- A segment of DNA containing the error is peeled away using helicase and removed. The gap is filled using the parental strand as the template with polymerase III and DNA ligase I, as described in DNA replication (3.3.1). The correct cytosine base is thus introduced. Mismatch repairs, which form the vast majority of the various types of DNA repairs available (see Section 4.6 in Chapter 4), reduce the frequency of natural errors 1000-fold. Nevertheless, errors in replication can evade detection. When an established error remains and is duplicated in subsequent cell cycles, it becomes a mutation. Replication errors are the main source of mutations, most of which are harmless, but a few are more serious and even deadly. Patients with colorectal tumors, for example, are defective in the mismatch repair system.

3.4 MITOSIS AND CYTOKINESIS

Mitosis and cytokinesis phases separate the interphases of a single cell and the two cells produced by duplication. The chromatin during mitosis is condensed and shaped, in contrast with the tangled mass of DNA in the interphase. In mitosis, a nucleus in a cell splits into two nuclei. This is followed shortly after by division of the parent cell into two daughter cells (cytokinesis), each of which contains the same nuclear content as the parent cell and approximately equal amounts of cyto-plasm. The mitosis mechanism ensures that the duplicated chromosomes, produced in the interphase, are divided evenly between two nuclei. As is the case with the interphase, mitosis is a continuous process but is conventionally broken up into four stages. They are preceded by the interphase stage and followed by cytokinesis, which is often included in a discussion of mitosis. Cytokinesis is followed by the next interphase. A representation of the whole process is shown in Figure 3.13. In simple terms, one function of mitosis is to segregate the two sets of chromosomes formed during interphase.

In addition to the spindle fibers (kinetochore microtubules), which attach to and allow the sister chromatids to split apart, there are other microtubules (asters, not shown), which connect the poles directly. During the four stages, there are a number of

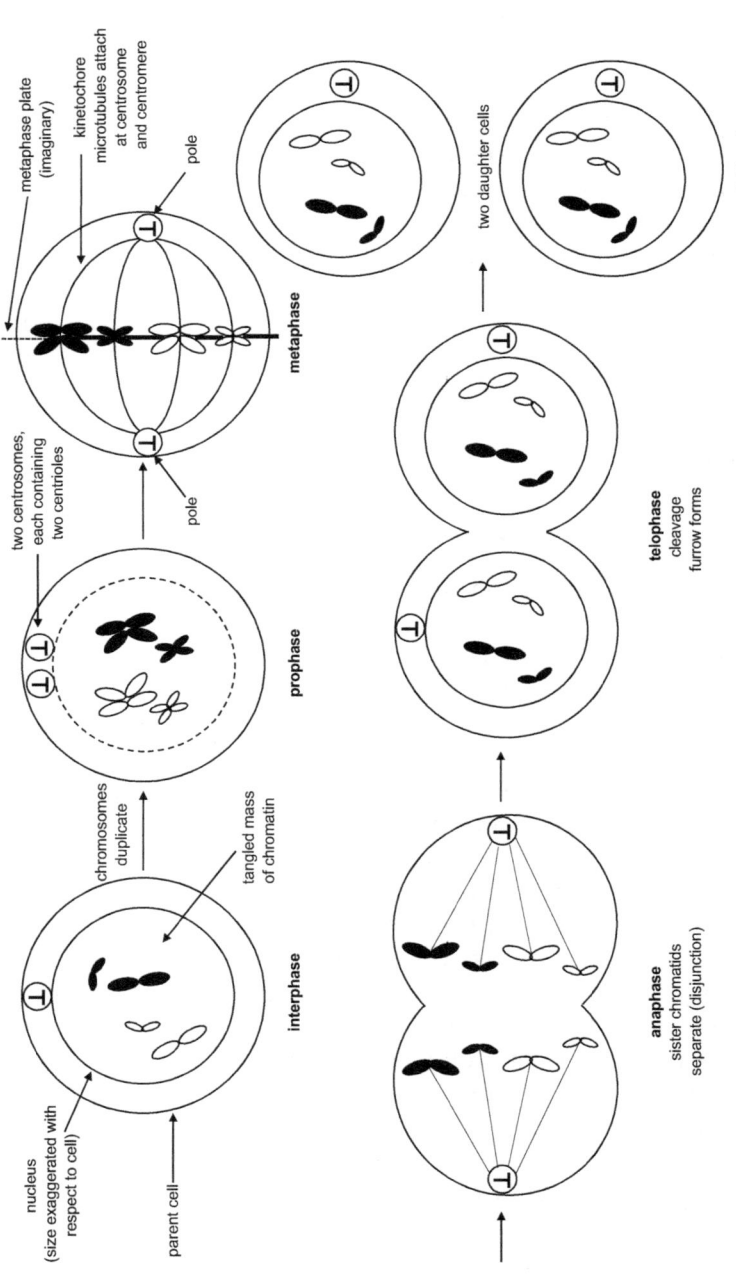

Figure 3.13 The sequence of events in mitosis and cytokinesis. Only two different chromosome pairs of the total set of chromosomes, *e.g.*, 23 pairs in humans (shown here in black and white), are shown with two different lengths. The longer could be human chromosome 1 and the shorter could be human chromosome 19.

happenings to the cell and contents, to the DNA (chromosomes) and to the cell machinery.

- **Interphase:** This is the major portion of the cell cycle with regards to duration and features cell growth. A distinct nucleus and nucleolus are visible with a microscope. DNA is replicated into two identical, loosely packed and thread-like chromatins, which are not clearly visible with a microscope. These are shown as chromosomes in Figure 3.13 to indicate their relationship with those in mitosis.
- **Prophase:** The nuclear envelope and nucleolus are disintegrating. DNA becomes more compact compared with that during interphase. DNA is packaged into a number of smaller and familiar X-shaped forms, consisting of two copies (maternal and paternal) of a pair of identical sister chromatids held together with a protein (cohesion) at the centromere. The mitotic spindle apparatus (Figure 3.13) begins to assemble. The two centrosomes (see Section 1.1.1) move towards opposite poles of the cell and spindle fibers gradually assemble between them. Cables of actin stretch from pole to pole of the mitotic spindle and, with the help of myosin, control the fibers' length and shape.
- **Metaphase:** The nuclear envelope breaks down and the nucleus has now disappeared. This now enables the spindle apparatus to fill the entire cell, be fully in place and have access to the chromosomes (Figure 3.13). Chromosomes further condense and approach the center of the cell. They line up on the metaphase plate, which is an imaginary equator that divides the cell into two. The line-up is in single file, an arrangement that allows the sister chromatids to attach to spindle fibers emanating from opposite poles (Figure 3.13). The single file arrangement ensures that there is no transfer of DNA between the homolog (maternal and paternal) chromosomes and ensures that the two cells will have identical chromosome content. This is in distinct contrast to the situation with meiosis, where a side-to-side arrangement of homologous chromosomes does allow the transfer of essential genetic material between them (see Section 5.1 in Chapter 5).

The two centrosomes are now at the polls. The fibers begin to contract and thereby pull each sister chromatid apart.

- **Anaphase:** The cell elongates and the first signs of its division appear. The sister chromatids split apart at the centromere (disjunction). The division of the paired sister chromatids at the metaphase plate into two separate chromatids (now termed chromosomes) is accomplished with the aid of the spindle apparatus. Shortening of the spindle fibers attached to the centrosome and chromatids ensures the breaking apart of the sister chromatids at the centromere and their transfer to the opposite poles of the cell. This will be close to the two nuclei that are going to form shortly. In most somatic animal cells, each spindle pole is focused on the centrosome. The spindle system remains intact.

- **Telophase:** The nuclear membrane is forming around two new nuclei at the two ends of the cell. The cytoplasm starts dividing as the two new cells become apparent. Two new sets of chromatids (more diffuse and uncoiling) arrive at opposite ends of the cell and are encircled by the two new nuclei. The spindle fibers disappear. The centrosomes are now outside the nuclei. The cleavage furrow forms. Telophase is largely a reversal of prophase, involving, respectively, DNA uncoiling and coiling, nuclei reappearing and disappearing, and spindle apparatus breaking and assembling. Mitosis is now over. It only remains for the process to be completed by cytokinesis.

- **Cytokinesis:** The division of the cell and its cytoplasm into two cells is completed in this phase. Membrane, cytoskeleton, organelles and the numerous associated proteins have been distributed to the daughter cells. A tangled mass of DNA similar to that in interphase is present in both cells containing identical genetic information. The contractile apparatus is disassembled and spindle fibers disappear. Beginning in anaphase, an inward wrinkling of the cell membrane at the equator ends forms. This is implemented by a contractile ring consisting of the proteins actin (mainly) and myosin. Cleavage furrow forms. The cleavage ring continues to become smaller until it causes a complete break and two cells are formed. The cleavage furrow bisects the cell at the equator, thus ensuring that cellular division results in equal sizes for the daughter cells and one nucleus per cell.

3.4.1 Departures from Normal Mitosis

Errors occurring in mitosis are amazingly rare. When they occur in early cellular divisions in the fertilized egg during pregnancy, the consequences may be particularly serious, leading, as they do, to children who will also carry any defect that the error may cause. There are two spots in particular where the mitotic sequence might be disturbed, leading in some cases to desirable results but more often to animal disorders.

3.4.2 Endoreplication

Endoreplication is a collective term for variations in the cell cycle in which DNA replication occurs in the interphase, but dissolution of the nuclear membrane and cytokinesis does not take place. The subsequent cell division is absent. The resultant polyploidy cells are usually larger and more active metabolically than normal cells. These polyploidy cells will contain more than the normal two complete sets ($2n$) of chromosomes. Sets of three chromosomes ($3n$) are termed triploid, sets of four chromosomes (*i.e.*, in humans, a total of 92 chromosomes in a nucleus) are called tetraploid and so on. Examples of polyploidy are more prevalent with animals such as fish, frogs and insects (as well as flowering plants). They are generally rare with mammals but can be found in cattle and mice.

- **Endomitosis:** The cell cycle may still show features of mitosis. The sister chromatids may still separate and retain their individuality but remain encapsulated in a single nucleus within a single cell. They occur in certain animal tissues, for example, in megakaryocytes. These are large cells (35–160 μm in diameter), which are present in bone marrow. Fragmentation yields large platelets, which are necessary for normal blood clotting. Megakaryocytes of humans normally exist in a range of polyploidies from $4n$ up to even $128n$ chromosomes (when it will have passed through seven S phases in the cell cycle). The mean ploidy number of human megakaryocytes is $16n$, which is lower than that of cats and dogs ($32n$–$64n$). These exist in one multi-lobed nucleus, large enough to be seen under a light microscope.

- **Endocycling:** Mitosis is bypassed altogether. In contrast to endomitosis, the sister chromatids do not separate but remain synapsed to give a cable-like polytene chromosome, which may contain many sister chromatids encased in an expanded nucleus and single cell. A striking example of these giant chromosomes, mainly found in insects, is displayed in the salivary gland cell of *Drosophila melanogaster* larvae that secrete the glue protein required for molting. Each of the four pairs of chromosomes in the fruit fly has undergone some 10 rounds of DNA replication, without any cell division and the two daughter chromosomes remaining attached throughout. The maternal and paternal homologs (as well as the sister chromatids) are aligned exactly so that there are $2 \times 2^{10} = 2048$ identical chromatid strands lying side-by-side to give four giant cables, which are large enough (about 1 mm) to be easily seen with even a low-power light microscope.

3.4.3 Aneuploidy

Most aneuploidy is caused by nondisjunction. This results from a failure of sister chromatids to separate as they normally do (disjunction) at the anaphase of mitosis. This may be the result, perhaps, of a faulty spindle apparatus (Figure 3.13). *Both* sister chromatids of one chromosome are transferred to one of the daughter cells at the expense of the other cell. Only one pair of the autosomes is involved. This leads, on division, to one cell containing $2n + 1$ chromosomes (trisomy) and one cell containing $2n - 1$ chromosomes (monosomy), rather than both cells containing the normal $2n$ complement (disomy; Figure 3.14). Nondisjunction may lead to mosaicism.

This phenomenon is not to be confused with polyploidy, where a whole set of somatic chromosomes undergo numerical change. Thus, for example, a human cell might contain 47 chromosomes in aneuploidy but 92 chromosomes in tetraploidy. Nondisjunction is much rarer in mitosis than in meiosis (see Section 5.2.3 in Chapter 5). It accounts for a few percent of persons with Down syndrome, but these usually have both normal and abnormal distribution of chromosomes in the cells to give a mosaic pattern.

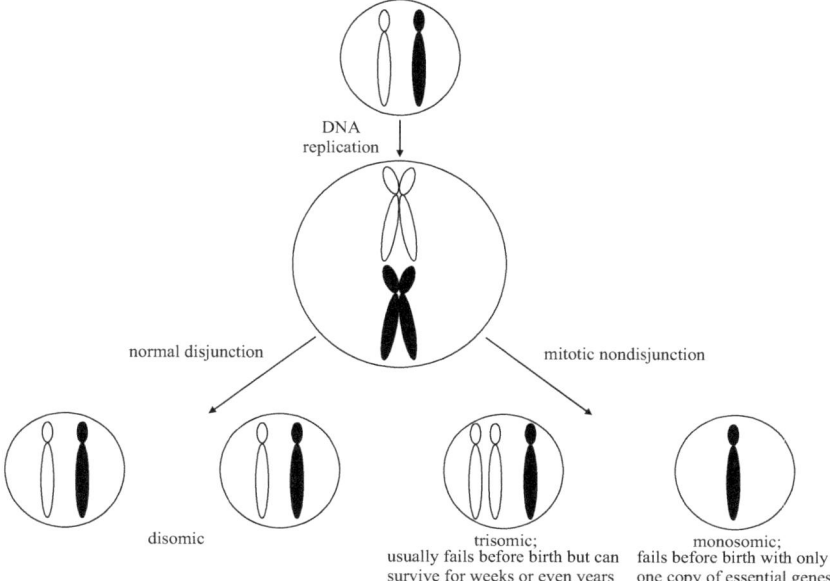

Figure 3.14 Normal disjunction and abnormal nondisjunction in mitosis, leading to normal disomy and an abnormal combination of trisomy and monosomy, respectively.

3.4.4 Mosaic Down Syndrome

Down syndrome is a classic example of a trisomy abnormality in human chromosome 21. Only 1–2% of persons affected by Down syndrome have the mosaic version as a result of mitotic non-disjunction. Ninety-five percent of sufferers result from meiosis nondisjunction (trisomy 21; Section 5.2.4 in Chapter 5).

Mosaicism arises when two different genotypes have developed from a single fertilized egg. The development of mosaicism involving chromosome 21 and leading to mosaic Down syndrome in humans is illustrated in Figure 3.15.

Mosaic Down syndrome occurs as an event in the early embryo. If nondisjunction were to occur in the first division in the zygote, all cells in the adult animal would be either monosomic or trisomic (Figure 3.14). Invariably, nondisjunction occurs in later cell divisions after the embryo has multiple types of cells (Figure 3.15). Only cells then originating from the improperly divided cell will be aneuploid. The rest of the cells dividing

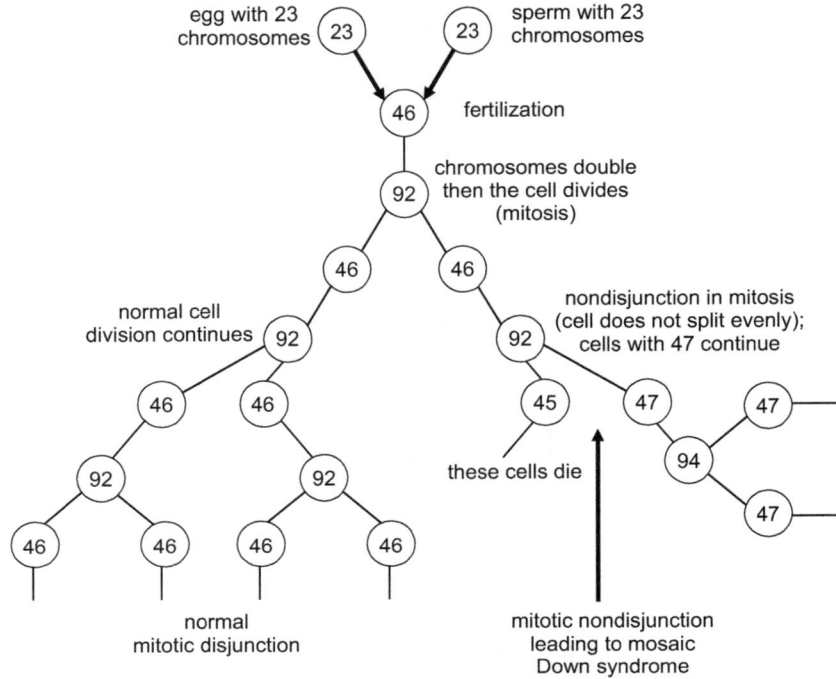

Figure 3.15   The formation of two types of cell, one with 46 chromosomes (usual) and one with 47 chromosomes leading to mosaic Down syndrome involving chromosome 21.

normally will be diploid cells in the developing baby. Because differentiation of cells has occurred, the percentage of cells that are mosaic may be different in different tissues of the body. In general, a lower degree of mosaicism (from fewer cells having the abnormal complement of chromosomes) is likely to lead to milder symptoms. This applies to mosaic Down syndrome, as well as the analogous disorders trisomy 18 and trisomy 13. These are lethal when all the cells have triple chromosomes 18 and 13, respectively, but have milder severity (physical features and longevity) in the mosaic versions. Only a few percent of people with the disease, however, have the mosaic versions, see Section 5.2.4 in Chapter 5. Mosaicism can affect any type of cell. Somatic mosaicism, as its name implies, affects somatic cells as, for example, in Down syndrome. Neurofibromatosis (Section 7.2.2 in Chapter 7) and Pallister–Killian (Section 4.4.5 in Chapter 4) are other diseases displaying this effect. Germline mosaicism on the other hand only arises in egg and sperm cells. A mutation usually occurs in early stem cells that give rise to the gonadal tissue. An important difference from somatic mosaicism is that it may be passed to offspring. Examples include Duchenne muscular dystrophy (Section 6.3.5 in Chapter 6), osteogenesis imperfecta (Section 7.2.5 in Chapter 7) and achondroplasia (Section 6.2.2 in Chapter 6). Mosaicism is more prevalent in autosomal dominant and X-linked disorders (Chapter 6). A somewhat different type of mosaicism results from X-inactivation (Section 7.1.3 in Chapter 7).

CHAPTER 4

DNA Mutations and Their Impact on Human and Animal Phenotypes

4.1 INTRODUCTION

This is a very important chapter encompassing all the varied changes that DNA can undergo and thereby produce faulty (mutant) genes and protein products. At the center of genetics, a mutation involves a functional change in the genetic material of a cell, specifically a change in the DNA or RNA sequences, and thus affects the sequence of amino acids in the resulting protein. If the change does not affect this sequence, it is a natural variant. A mutation is not necessarily irreversible, but if it involves somatic cells (acquired mutations), these will disappear when the owner dies. However, if the mutation involves the reproductive cells (germline mutation), an offspring may inherit the mutation in all its cells and have a hereditary ailment (inherited mutation). If after conception, growth and development of an embryo, only some cells have the mutation, mosaicism results.

The frequency of mutations is rare and may be inherited or acquired during a person's lifetime. Only a very small percentage of mutations cause genetic disorders, usually by altering or removing a critical encoded protein. The animal affected by the

Animal Genetics for Chemists
By Ralph G. Wilkins
© Ralph G. Wilkins 2017
Published by the Royal Society of Chemistry, www.rsc.org

mutation (the mutant) may have discernible changes in the resulting phenotype, that is in its appearance, behavior or traits (disorders or diseases) compared with the wild-type animal, which has the phenotype most common in nature. The intensity of the phenotype change is likely to depend on the magnitude of the mutation, namely how much the DNA is changed or the chromosome involved. The mutation may arise as a random event during the formation of the egg or sperm, or in the early embryo (see Section 5.3 in Chapter 5), in which case there will have been no history of the disease in the family. Alternatively, it may be inherited from an affected parent.

4.2 TYPES OF MUTATION

There is a wide range of types of alterations in DNA that lead to a mutation. Slight changes, involving only one or a few nucleotides, can only be detected by full sequencing techniques. On the other hand, whole chunks of DNA, involving millions of base pairs and even entire genes, may be lost or moved around within a chromosome or between chromosomes. These may be detected using karyotype patterns or by spectroscopic techniques (see FISH and SKY in Section 1.4.3 in Chapter 1).

4.3 SMALL CHANGES IN DNA

4.3.1 Single Nucleotide Polymorphism

A single nucleotide variation or point mutation in either the coding or non-coding region of the gene (single nucleotide polymorphism, SNP, pronounced "snips") is the most abundant type of genetic variation in the human genome and accounts for 90% of the differences between individuals. SNPs occur about one in every 300 nucleotides in the human genome. There are a variety of SNPs, most of which are harmless. Those within the intron regions (see Section 3.1 in Chapter 3) are termed intronic (iSNP) and those located between genes, *i.e.*, the intergenic (gSNP) regions, are much more plentiful than the more harmful ones that are within the coding (cSNP) and regulatory (rSNP) regions. About 70% of SNPs involve a nucleotide replacement of C by T. For the altered sequences to be considered a significant

SNP, they must occur in over 1% in at least one population. SNPs change little through generations.

- The single nucleotide change in, for example, GGC to GGA will not lead to a change in the encoded protein since both codons lead to one particular amino acid, namely glycine (Figure 3.6). Such mutations are termed silent mutations.
- A single nucleotide change may, however, result in a different amino acid in the translated polypeptide chain and protein. This protein may still function, although there may be less of it than normal. Alternatively, the protein may not even function, or may have a new function. Such a change is termed a missense mutation.
- In a nonsense mutation, a sense codon may be converted into a premature chain-terminating stop codon. This results in a truncated protein product, which may or may not be functional. Examples of missense mutations occur with bobtail dogs and myotonic goats, and of nonsense mutations with harness racing trotters and the human McArdle disease.

4.3.2 Missense Mutations

ATC (Isoleucine) ⇒ ATG (Methionine)

This change occurs in the first exon of the *Brachyury* (*T*) gene on the dog chromosome 1q, which encodes protein T, required for mesoderm (muscle) formation and differentiation. The mutated protein T fails to bind to the T site necessary for correct muscle function. This mutation leads to a shortened or missing tail (bobtail). It is present in the Pembroke Welsh Corgi, as well as a large number of other breeds of dogs. One copy of the mutant gene is sufficient to cause the bobtail, while two copies is lethal (Section 6.2.4 in Chapter 6).

GCC (Alanine) ⇒ CCC (Proline)

This change occurs in the *chloride channel 1* (CLCN1) gene on goat chromosome 4, which encodes a skeletal muscle channel protein. This allows chloride flow and coordinates muscle contraction and relaxation. In the mutated protein, the channel opening is delayed and muscle contraction is prolonged. This mutation causes myotonia congenita. The myotonic goat may stiffen and sometimes fall over when startled. One copy of the mutant gene is sufficient to cause the effect. The gene and its faulty variant is known in a variety of animals, including humans. Most of the mutations in the gene in human chromosome 7q result in Becker disease.

4.3.3   Nonsense Mutations

TCG (Serine) ⇒ TAG (Stop)

This change occurs in the *DMRT3* gene on horse chromosome 23, which encodes a DNA-binding protein in nerve cells in the spinal cord, crucial in coordinating leg movement in the horse. In the mutant protein, the premature stop leads to a protein that is missing 172 amino acids at the end of the normal protein, which has 472 amino acids. In spite of this large, nonsense mutation, the horse still moves, but the mutation leads to a novel gait displayed by harness racing trotters. These horses trot or pace at high speed with the legs on one side moving forward at the same time. A copy of the mutation is required in both chromosomes.

CGA (Arginine) ⇒ TGA (Stop)

This change occurs in the *phosphorylase, glycogen, muscle* (PYGM) gene on human chromosome 11q. The gene encodes the myophosphorylase enzyme, which breaks down glycogen to glucose in muscles. The latter is used to provide energy during exercise. This common mutation leads to the production of decreased amounts of enzyme, causing severe muscle pain and cramp during exercise. The disorder, McArdle disease, is very rare, but it is one of the most common disorders of muscle metabolism. A copy of the mutant gene is required in both chromosomes (recessive mutation; Section 6.2.10 in Chapter 6). (N.B. Medical geneticists use the non-possessive form for the designation of a disease named after an individual.) Some 15–30% of inherited diseases, including cystic fibrosis, hemophilia, Duchenne muscular dystrophy, aniridia and retinitis pigmentosa, are as a result of a nonsense mutation.

4.3.4   Insertions and Deletions (INDEL) Mutations

Insertion and deletion of DNA material (an INDEL mutation) may involve from 1 to 1000 base pairs and be either terminal or interstitial (Section 1.5.1 in Chapter 1). Detection of such mutations, as with the case of point mutations, requires sequencing techniques. If the number of nucleotides involved is three (or a multiple thereof), only one (or more) amino acid is lost or added since each amino acid requires a codon of three nucleotides (Section 3.1.5 in Chapter 3). If, however, the number of nucleotides involved is a non-multiple of three, the disruption is termed a frameshift mutation, and a new in-frame and a different amino acid sequence results (see Section 4.3.6). A truncated, nonfunctional form of a protein usually results. Examples of mutations without and with frameshift are shown in two human diseases.

4.3.5   Small Three Nucleotides Deletion

ATC AT[CTT]T GGT ⇒ *ATC ATT GGT*

Ile Ile Phe Gly ⇒ **Ile Ile Gly.** This change occurs in the *cystic fibrosis transmembrane conductance regulator* (*CFTR*) gene (6000 nucleotides) on human chromosome 7q. The gene encodes a CFTR protein, which is a chloride channel and essential for the correct functioning of lungs and other organs.

In the mutation, a [CTT] triplet (Leu) is deleted, which is a simple amino acid deletion not involving a frameshift. The mutated protein is only one amino acid (Phe, symbol F) shorter than the normal protein but is misfolded, enzymatically degraded and not able to form a channel. Chloride ion transport is aborted, causing sodium and chloride ion imbalance, and this is associated with a build-up of thick, sticky mucus in the airways, often leading to respiratory infections. This delta F508 mutation occurs in about 70% of cystic fibrosis sufferers and is the most common of more than 1800 mutations that have been recorded for this gene. Two gene copies are needed but different severities are observed (variable expressivity; see Section 7.2.1 in Chapter 7).

4.3.6 Small Four Nucleotides Insertion

ATA TC*C TAT GCC CCT GAC ⇒ ATA TCT ATC CTA TGC CCC TGA

Ile Ser Tyr Ala Pro Asp ⇒ Ile Ser Ile Leu Cys Pro Stop. Present in the *hexosaminidase A (alpha polypeptide)* (*HEXA*) gene on human chromosome 15q, the gene encodes one part of an enzyme (α subunit) that combines with a β subunit (encoded by the beta-*HEXB* gene) to form a functioning enzyme: hexosaminidase A. This breaks down lipids in nerve cells in the brain and spinal cord, particularly in the fetus and young children. In the mutation, four nucleotides (TATC) are inserted at *. This leads to a quite different sequence of amino acids after the insertion (frameshift) and a premature stop codon (TGA). A single copy of the mutation leads to reduced amounts (50%) of enzyme, although still sufficient for breaking down fatty waste products. Two copies of the mutation result in no functional enzyme. It is the most common of a number of different types of mutation for the Tay–Sachs condition, which results in paralysis, blindness and, usually, an early death. The two mutant copies necessary may be a combination of two different types of the 100 or so mutations known. It is a rare example of a disorder controlled by a single gene (Figure 1.13). The disease is largely contained by genetic counseling in vulnerable communities (primarily Ashkenazi Jewish).

4.3.7 Cystic Fibrosis: Classes of Mutations

The different types of mutations discussed so far, and where they impact in the cell, are well illustrated by the disease of cystic fibrosis. Different types of mutation of the *CFTR* gene (Section 4.3.5) can give rise to cystic fibrosis. The nature of the mutation will determine the fate of any encoded mutated CFTR protein, as well as the severity of the disease. Of paramount importance is how the passage of the normal protein to the cell membrane (where it would normally form a functioning channel; Figure 4.1) is modified in the products arising from the various mutations or indeed if they reach the membrane at all.

Normally, the CFTR DNA is converted to CFTR mRNA in the nucleus and then becomes nascent CFTR protein in the rough

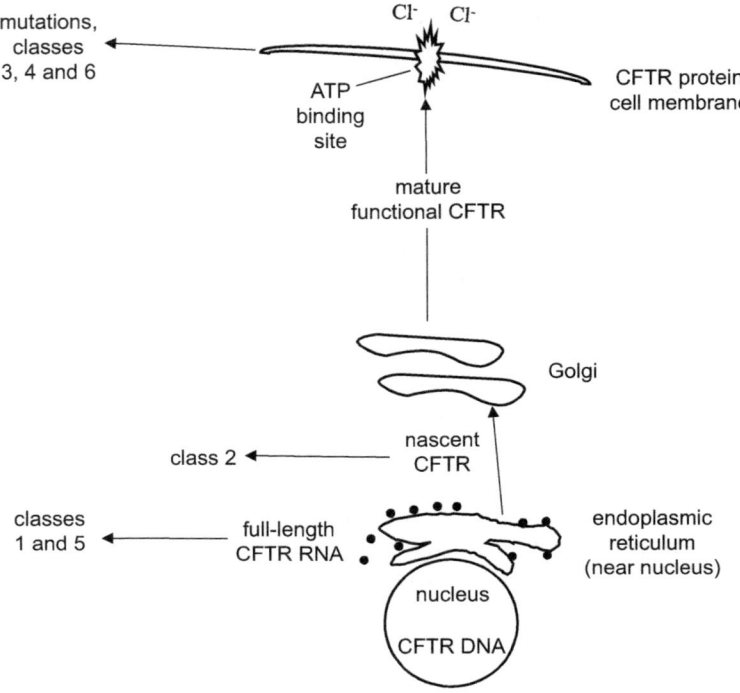

Figure 4.1   The normal route of the CFTR protein in progressing from its site
of formation to the cell membrane, showing the locations where
the various classes of mutation make their effects felt.

endoplasmic reticulum. It is glycosylated in the Golgi apparatus
during post-translational processing to give the mature, functional CFTR. It is then sent to various parts of the cell, including
the secretary vesicles. It becomes the fully functional CFTR
chloride ion channel in the plasma membrane. It has been decided that there are six classes of mutation. With these mutations either no protein reaches the cell membrane and no
channel is created or protein material reaches the membrane
but, for various reasons, fails to produce an operational channel
(Figure 4.1). The classes and one important mutation for each
are:

- **Class 1.** A truncated RNA results from a nonsense mutation:
  G542stop. No protein is synthesized, and therefore no
  channel is created.

- **Class 2.** Full-length RNA is obtained but with a misfolded protein and no channel is created. It is the most common mutation (Section 4.3.5).
- **Class 3.** Full-length RNA is obtained, but an abnormal channel with a defect in the channel opening results from a missense mutation (G551D).
- **Class 4.** A defective channel protein results from the missense mutation R117H. The protein reaches the membrane, but ion conductivity is altered and the channel fails to function correctly.
- **Class 5.** Both correct and "incorrect" mature RNA results from a missense A445E mutation. Insufficient functional CFTR protein results from this second most common mutation.
- **Class 6.** A channel protein is produced but is rapidly degraded. It results from a very rare nonsense mutation: Q1412stop.

Knowledge of these types of mutation is invaluable in the design of various drugs to focus on the mutation and combat the disease, see Section 8.5.1 in Chapter 8.

4.4   LARGISH AND LARGE CHANGES IN DNA

Many mutations arise from large chromosomal changes, which can be diagnosed by SKY analysis (Section 1.4.3 in Chapter 1). The types are shown in Figure 4.2.

4.4.1   Duplication

Larger insertion mutations may involve duplication of not just one copy of the same tri- or tetra-nucleotides (Section 4.3.6) but very many being inserted in either the coding or the non-coding part of the gene.

Polyglutamine disorders involve CAG repeats (CAG codes for glutamine). Nine have been described, of which Huntington disease is the most familiar.

Huntington Disease

The *huntingtin* (*HTT*) gene on human chromosome 4p encodes huntingtin, a protein whose function is unknown, although it is

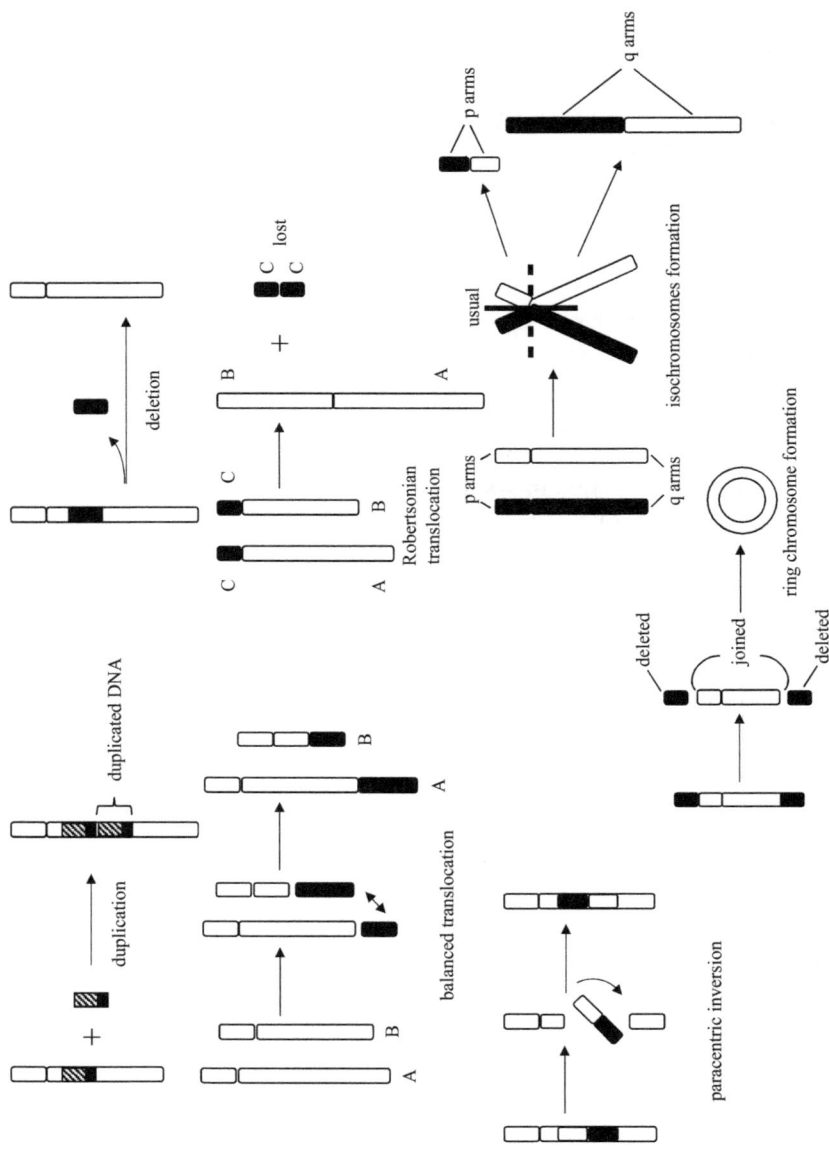

Figure 4.2   Types of large chromosomal changes.

almost certain that it plays an important role in the action of brain neurons. In the *HTT* mutation that causes Huntington disease, 10–35 repeats of the CAG triplet for the normal gene are expanded to 36–140 repeats in the mutant form. This expansion occurs in the coding region, producing an abnormally long version of the protein, which breaks into small, toxic clumps. These clumps disrupt neuronal action. Huntington disease is associated with progressive brain disorder and terrible associated symptoms. The onset is mostly in adulthood. The size of the repeats may increase from one generation to the next (to be avoided if possible), leading to an earlier onset of the disease. Only one copy of the mutation is necessary to invoke the disease.

Trinucleotide repeat expansion (non-polyglutamine) disorders involve another three nucleotide combinations. In these, gene expression is switched off in different ways and in different parts of a gene. In addition, the intensities of the diseases usually depend on the degree of the repeat expansion (see Section 7.2.6 in Chapter 7). Three human diseases illustrate these points.

Fragile X Syndrome

The fragile X mental retardation 1 (FMR1) gene on the Xq chromosome encodes fragile X mental retardational protein (FMRP), which plays an important role in relaying nerve impulses. FMRP carries a number of different RNAs around the cell, which regulate the production of other proteins. Mutation leads to the loss of regulation of these RNAs. Mutations in the gene on the Xq chromosome cause nearly all the cases of the syndrome, which include a range of developmental problems including learning disabilities. It more severely affects males, with resulting unusual physical characteristics. It is second to Down syndrome for inherited intellectual disability in males. The mutations arise from an abnormal number of CGG repeats in the 5′ UTR region of the gene (Section 3.1 in Chapter 3); although a non-coding part, they still interfere with its function. Normal DNA has 6–54 copies of the trinucleotide at the relevant locus. The full X-linked dominant mutation (see Section 6.3.1 in Chapter 6) involves more than 200 repeats.

Friedreich Ataxia

The *frataxin* (*FXN*) gene encodes frataxin protein and is located on chromosome 9q. The protein plays a major role in the biogenesis of enzymes involved in the respiratory chain in the mitochondria. The most common mutation involves an expansion of a normal 5–33 GAA repeat to 66–1700 repeats in intron 1 of the gene. The increase disrupts the production of frataxin and leads to accumulation of mitochondrial toxic iron. The mutation leads to nervous system and muscle damage, a clumsy gait, slurred speech and general muscle weakness. It is the most frequent heritable ataxia in Europe, with 1 in 25 000 suffering with the condition. Most of these are homozygous for this autosomal recessive condition (Section 6.2.10 in Chapter 6).

Myotonic Dystrophy

The *dystrophia myotonica protein kinase* (*DMPK*) gene encodes myotonic dystrophy protein kinase, which has an unclear function, on human chromosome 19q. It does appear to play an important role in communication within heart and brain cells. Myotonic dystrophy type 1 is an autosomal dominant condition (Section 6.2.1 in Chapter 6) caused by mutations in the 3' UTR region (Section 3.1 in Chapter 3) of the gene, so the entire coding region, as well as the expansion, can be copied into the mRNA. Normally, there are 3–30 CTG repeats in this region. Expansion to 50–4000 repeats results in an altered mRNA. This traps proteins to form clumps within the cell, which prevent normal muscle cell functioning, leading to progressive muscular weakness and prolonged muscle contractions (myotonia). The classic adult type is present in 1 in 8000 of the population and is the most frequent of all muscular dystrophies. A differentially methylated region near the CTGs undergoes hypermethylation (Section 7.4.2 in Chapter 7) as a function of expansion size, with reduced expression of a neighboring gene. The significance is at present uncertain.

A disease arising from duplication of a large segment of DNA is Charcot–Marie–Tooth (Section 5.1.1 in Chapter 5).

4.4.2 Deletion

Large deletion mutations (Figure 4.2) can be identified by SKY analysis of the amniotic fluid from the fetus, as well as adults. The loss of millions of base pairs, and therefore very many associated genes with their genetic information in various human chromosomes, is associated with some rare syndromes, which are diseases with a group of symptoms. They include:

- DiGeorge syndrome (22q-): Inability to fight infections.
- Jacobsen syndrome (11q-): Mental retardation and multiple birth defects.
- Cri-du-Chat syndrome (5p-): Facial muscle and intellectual problems.

Again, the severity of the diseases often depends on the size of the deletion. Most cases of these diseases arise from random events during the formation of the sperm or egg, or in early fetus development. Affected persons therefore rarely have a history of the disorder in their family.

4.4.3 Translocations: Chronic Myeloid Leukemia

Balanced (reciprocal) translocation, the most common translocation, is the exchange of large portions of chromosome between non-homologous chromosomes (Figure 4.2). Each chromosome therefore gets part of the other chromosome. An even exchange of genetic material means there is no loss or gain of total genetic information and so usually no abnormality results. However, balanced chromosome translocations are considered pathogenic events in hematopoietic (pertaining to the formation of red blood cells) malignancies. An example of this is chronic myeloid leukemia (CML) in humans. In CML, a reciprocal translocation between chromosome 9 and chromosome 22 occurs, giving an extra-long chromosome 9 and an extra short chromosome 22 (Philadelphia chromosome), see Figure 4.3.

Proto-oncogene *ABL1* exists on chromosome 9q. A proto-oncogene is a normal gene that can become an oncogene (which induces uncontrolled cell proliferation) because of a mutation or

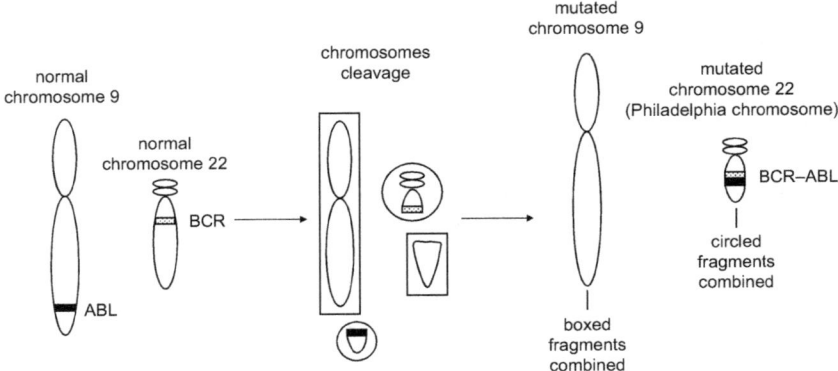

Figure 4.3 Reciprocal translocation involving human chromosomes 9 and 22.

increased expression. The proto-oncogene encodes tyrosine kinase, an enzyme implicated in a number of cell properties.
During the translocation, most of the *ABL* gene on chromosome
9 ends up fused head-to-tail in the middle of a gene called
breakpoint cluster region (BCR), which appears on chromosome
22q. The hybrid gene *BCR–ABL* encodes an excess of tyrosine
kinase. Certain blood cells grow out of control, produce a large
excess of white blood cells and promote CML. A person may
possess the Philadelphia chromosome for many years until a
"blast crisis" is initiated for unknown reasons and fully fledged
chronic myeloid leukemia results. This is one of the few cancers
caused by a single specific genetic mutation and this has contributed to the success of the drug in the treatment of the disease
(see Section 8.5.2 in Chapter 8).

Robertsonian Translocation

In a Robertsonian translocation (Figure 4.2), two acrocentric
chromosomes get stuck to each other. The much longer q-arms of
the chromosomes fuse at a centromere and give rise to one larger
hybrid chromosome, containing the two q-arms. The p-arms
contain few genes and are lost. The hybrid will therefore contain
most of the genes of the two contributing chromosomes. About
4% of Down syndrome children are due to a Robertsonian
translocation before or at conception, when an additional
chromosome 21 is fused with another acrocentric chromosome,

usually 14. The other causes of Down syndrome are detailed in Section 3.4.4 in Chapter 3 and Section 5.2.4 in Chapter 5.

Cattle

Cattle normally have 60 chromosomes (58 + XX or XY). All the autosomes have centromeres near the end of the chromosome. The most common type of Robertsonian translocation is 1/29, with a fusion of the largest cattle chromosome with the smallest chromosome. The hybrid is most common in European breeds. Such cattle have only 59 chromosomes, but their phenotypes are normal. They do, however, have fertility problems.

4.4.4   Inversion

In an inversion, two breaks are made in one chromosome, a piece of DNA is removed, turned around and reinserted into the chromosome. The breaks are then rejoined (Figure 4.2). The sequence of genes is reversed. Balanced inversions can involve the centromere, when both the short and long arms are involved (pericentric inversion). If only one arm is involved, the more common paracentric inversion results. If, as is often the case, there is no net loss of genetic material, usually no phenotype abnormality is shown unless breaks are within essential genes (see the tobiano horse, discussed later in this section). In unbalanced inversion, genes may have been deleted or duplicated and now there is an increased likelihood of defects arising from the inversion. As with most large chromosomal changes, the extent of any disorder depends on the size of the change, in this case its location and the type of inversion. Examples of inversion from the human and equine kingdoms are shown in hemophilia A and by the tobiano horse.

Hemophilia A

This disease is associated with the *coagulation factor VIII (F8)* gene very near the tip of the q-arm of the X chromosome, which encodes coagulation factor VIII (or 8). This protein, when activated by blood vessel injury, becomes part of a cascade of chemical reactions responsible for blood clotting. About 50% of sufferers of a severe form of hemophilia A have an inversion of a large segment involving intron 22 in the *F8* gene. This leads to complete disruption of the normal gene. This results in either an abnormal version or more than 99% reduction in the normal amount of coagulation factor VIII and leads to continuous bleeding after a mild injury. About 1 in 5000 males worldwide are affected. It is an X-linked recessive disease (Section 6.3.4 in Chapter 6) and largely inherited. Milder hemophilia symptoms are usually associated with point mutations. Slightly less common is hemophilia B, arising from a mutation in a related gene, *F9*, which afflicted the royal families of Europe (Section 6.3.5 in Chapter 6).

Tobiano Horse

The *KIT* gene on chromosome 3q of the domestic horse encodes a large protein called tyrosine kinase receptor. This is involved in many cell processes during embryonic development and is responsible for many different white patterns on horses. In the tobiano horse, there is a large chromosomal inversion about 100 kb downstream from the *KIT* gene on horse chromosome 3 (ECA3; Figure 4.4). It is detected by FISH analysis. A paracentric inversion may disrupt regulatory sequences for the *KIT* gene and cause the characteristic pattern found only on tobiano horses, which include Shetland ponies and many other breeds.

The head is usually colored and the horse can have large white patches on the leg and elsewhere, with a usually brown or black body. Expressivity (a range of phenotypes; Section 7.2.1 in Chapter 7) is prevalent. The mutation is also present in a number of other animals, including mice, where it has been studied extensively. The same mutation of the *KIT* gene on human chromosome 4 causes a piebald condition, in which patches of the skin lack pigmentation and appear white.

4.4.5 Isochromosomes

A normal human chromosome consists of one short (p) arm and one long (q) arm. An isochromosome has either two p-arms or two q-arms. The isochromosome is produced during anaphase of

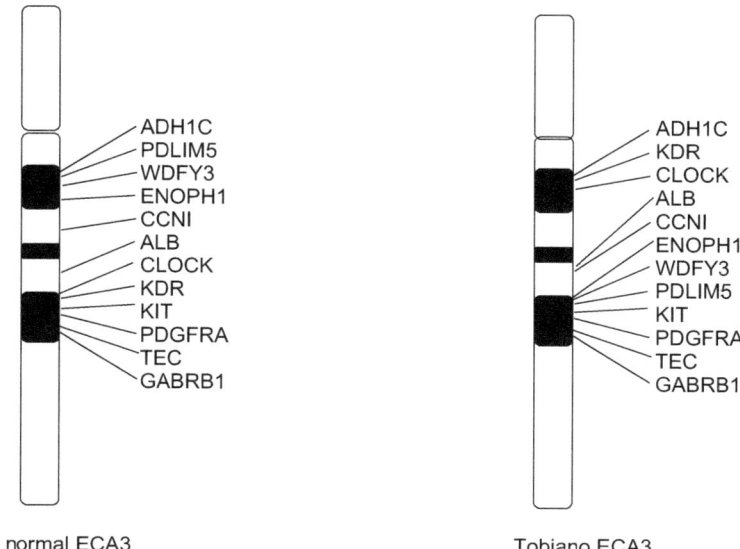

Figure 4.4 Normal horse (*Equus caballus*; chromosome 3, ECA3) and DNA inversion in the Tobiano horse. Note the reversal of genes.

mitosis or meiosis II, when there is transverse rather than the usual separation of the sister chromatids (Figure 4.2). The two arms of the isochromosome are identical in length and content but are mirror images of each other.

Pallister–Killian Mosaic Syndrome

People with the rare Pallister–Killian mosaic syndrome will have some cells with the usual two human chromosomes (12) and other cells with these two chromosomes plus one copy of an isochromosome, likely produced from nondisjunction. In this case, the isochromosome has two p-arms and no q-arms, so the abnormal cell will have a total of four p-arms, and therefore four copies of the genes in the p-arm of chromosome 12. An alternative name for the syndrome is therefore tetrasomy 12p. The unusual genetic material in some of the cells disrupts normal development, leading to life-threatening birth defects and serious problems thereafter (mental disturbances, muscular problems, *etc.*). If there were no cells with the normal chromosomal content, it is very likely that the individual would not survive. The extra isochromosome with the two 2p arms shows up nicely in the karyotype.

4.4.6   Ring Chromosome

A ring chromosome, which is rare, arises when the two telomere ends of a chromosome fuse to give a circular structure (Figure 4.2). The fusion may occur with or without loss of any material. Any loss in forming the ring may mean the loss of critical genes at the tips of the chromosome, such as those important in growth and development. Only a few human chromosomes, for example, 14 and 20, give rise to ring chromosomes. They invariably occur as a random event in the sperm or the egg. One copy of the ring chromosome and one copy of the normal chromosome (mosaicism) distributed in the cells is fairly common.

Turner Syndrome

About 40% of the cases of Turner syndrome arise from the loss of one of the two X chromosomes of the female (Section 5.2.4 in Chapter 5), of which about 30% have the mosaic version. Turner syndrome may also occur if one of the X chromosomes is not missing but has a structural abnormality. This includes iso-chromosome Xq (15% of all cases) and ring chromosome X (5%). Apparently, the complete or partial loss of important genes in the q-arm of the X chromosome give rise to the disorder.

4.5   RATES OF MUTATIONS

The mutation rate can be defined as the rate at which various types of mutation occur over time. It is difficult to assign numbers, but they can be considered qualitatively:

- Some regions of the genome are much more susceptible to mutations, even by as much as 1000-fold. The nature of the gene, differences in repair mechanisms and susceptibility to outside factors, such as mutagens (chemicals and X-ray), are all contributing factors.

- Mutation rates vary with the animal. One of the most important influences on the mutation rate is the generation time (GT) of the animal. This is the usual interval of time between an animal's birth and the birth of the first offspring. The more generations an animal has per unit time, the more chance for mutational errors to creep in, and therefore a higher rate of mutations. For example, wild mice (GT 6 months) have a higher rate of mutations than humans (GT 20 years).
- A higher mutation rate for mammalian males than females has been consistently observed. The difference has been assessed as anywhere between a factor of two and eight, depending upon the experimentalists and the loci examined. In males, cell divisions are continuous and many divisions have taken place before sperm is produced. In contrast, female germ cell divisions terminate at birth and meiosis is completed only when the egg matures. There is, therefore, a much greater number of cell divisions in the male than the female germlines and this offers a plausible explanation for a higher mutation rate in males. A recent study analyzed the STR (Section 1.5.1 in Chapter 1) in the DNA of some 24 000 parents and their children from three continents. STRs only mutate during cellular replication. It was found that the overall DNA mutation for fathers is seven times higher than for mothers, who had a low and lifelong constant mutation rate.

4.6   CAUSES AND REPAIR OF MUTATIONS

Mutations can arise from mistakes made during the various operations that occur in normal cellular activities. These include transcription and translation in gene expression, DNA replication and the cell cycle, especially involving mitosis and meiosis. They are spontaneous and their repair has to be addressed. Whereas replication mismatch repair addresses replication errors (Section 3.3.3 in Chapter 3), base and nucleotide excision repairs are mainly concerned with chemical damage to the DNA. These are probably the three major DNA repair systems.

4.6.1 Base Excision Repair

This counteracts the natural damage that DNA can undergo during the cell cycle. Base lesions arise from hydrolytic, oxidative and alkylation damage to any of the four nucleotides. Endogenous oxidants, such as $^{\bullet}$OH radicals and hydrogen peroxide, can alter a nucleotide, resulting in structural changes in the DNA and resulting mutations. These types of mutation do not distort the helix. Base excision repairs the single strand breaks. The simple description disguises the complexity of this and other repair mechanisms:

- A lesion-specific DNA glycosylase recognizes and removes an incorrect base. For example, the nucleotide cytosine can easily lose an $-NH_2$ group by hydrolysis and be converted to uracil. This cannot base-pair with guanine, unlike cytosine. The removal leaves a deoxyribose sugar lacking its base (an abasic site).
- An endonuclease and phosphodiesterase removes the remainder of the nucleotide.
- DNA polymerase restores the correct nucleotide and DNA ligase seals the single strand break.

No human disorders caused by inherited repair deficiencies have been observed, but base excision repair protects against neurodegeneration, cancer and aging.

4.6.2 Nucleotide Excision Repair

This process is similar in principle to base excision repair but is usually invoked when large distortions occur. Nucleotides can be changed in a number of ways by externally generated chemicals. Chemicals in smoke, for example, can cause mutations by simulating nucleotides and altering the shape of a segment of DNA. 5-Bromouracil is a synthetic analog of thymine (a Br group replacing an H; Figure 1.5) and can replace thymine during replication. UV light can induce a C=C double bond between two adjacent thymine nucleotides, and thereby distort the sugar–phosphate backbone and change normal replication and transcription. In nucleotide excision repair in humans, an oligonucleotide encompassing the damaged area is removed and replaced in several steps:

1. The abnormal structure is recognized by a multienzyme complex.
2. Endonucleases promote incision on both the 5′ and 3′ sides of the damage.
3. A DNA helicase removes the segment encompassing the damage lying between the two incisions, leaving a gap of 24–32 nucleotides.
4. The gap is repaired using the undamaged strand as a template with DNA polymerase and ligase.

Some people lack one of the enzymes involved in the repair process that removes sunlight-induced DNA lesions. Such patients suffer from xeroderma pigmentosum and have skin extremely sensitive to UV damage and potential skin cancers. In many cancers, one or more of the DNA repair types may be partially or completely switched off. This makes the cancer cell very dependent on the unaffected repair systems. If this is also suppressed, cancer growth will be slowed down or even stopped, obviously very desirable. Cancer drugs that inhibit this repair system are being actively sought (see Section 8.5.1 in Chapter 8).

4.6.3 Double Strand Break Repair

Double strand breaks (DSBs) may arise from ionizing radiation and certain chemical mutagens. They may also be produced deliberately in gene editing, where the gene may undergo knock out or replacement (see CRISPR/Cas9 in Section 8.4.10 in Chapter 8). DSBs are among the most pernicious types of DNA damage and, if unrepaired, may cause serious chromosome abnormalities and even cell death. Since both strands of the DNA are broken, there is no strand available to act as a template, unlike the repairs discussed in the previous sections. The two main repair mechanisms available are, in brief:

- Homology directed repair (HDR), which is involved in the S and G2 phases of the cell cycle (see Section 3.2 in Chapter 3) and is largely error-free since it relies on an undamaged sister chromatid or homologous chromosome (available after the S phase) as a template. The damaged DNA is replaced with undamaged DNA in a very complex process.

- Non-homologous end joining (NHEJ) takes place at the same time as HDR and, in principle, is a simpler process. It is involved in all phases of the cell cycle. The DSBs are repaired by attaching specific proteins to each free end, which draw them with the aid of other proteins sufficiently close to rebind the ends with DNA ligase. This is error-prone and bases may be lost or gained in the process.

CHAPTER 5

The Generation and Bringing Together of the Sex Chromosomes

5.1 MEIOSIS

In animals, meiosis only occurs when gametes are formed in the reproductive organs, namely the ovaries in the female and the testes in the male. Meiosis produces the sex cells (gametes), *i.e.*, the egg (ovum) in females and the sperm in males. The parent female or male diploid cell undergoes one round of DNA replication in interphase (much as in mitosis). This is followed by two successive nuclear divisions, with corresponding cell divisions (recall there is only one nuclear and one cellular division in mitosis). As a result of these sequences (meiosis I and II), four daughter cells (haploid gametes) result, each containing half the number of chromosomes as there is in the parent cell. This is illustrated in Figure 5.1.

In meiosis I, homologous chromosomes are separated (primary disjunction). Meiosis II much more resembles mitosis, as it is now that sister chromatids are parted (secondary disjunction) and four haploid cells result (Figure 5.1).

Meiosis I and II are subdivided into phases, whose designations are the same as, and resemble, those in mitosis. There are, however, major differences, particularly in meiosis I.

Animal Genetics for Chemists
By Ralph G. Wilkins
© Ralph G. Wilkins 2017
Published by the Royal Society of Chemistry, www.rsc.org

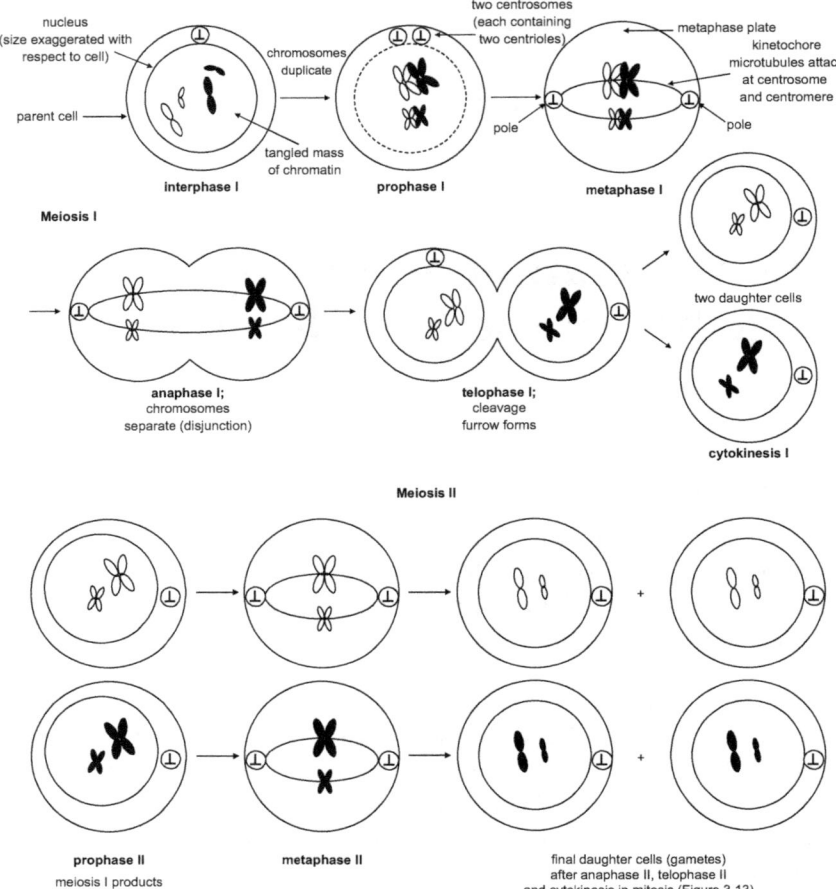

Figure 5.1   Representation of the transformation by meiosis of one diploid cell
with a small set of two chromosomes into four haploid cells, each
with a set of one chromosome. Only two different pairs of chromosomes, one larger and one smaller, of the 23 pairs in humans are
depicted. Their parental origins are in black and white. One
chromosome pair could also line up so that a black designated
chromosome leaves the metaphase I plate with a white designated
chromosome. In this case, the final four gametes will each contain a
black and a white chromosome pair (see Figure 6.6).

The happenings in the cell and cell machinery in meiosis I and II
are very much as in mitosis.

- **Interphase I:** This is similar to the same phase in
  mitosis. Chromosomes are doubled. A centrosome with

two centrioles is duplicated. There is no spindle yet in the cell.

- **Prophase I:** This phase occupies around 90% of the total time spent in meiosis. The cell remains intact, but the nuclear envelope and nucleolus begin to disappear by the end of the phase. DNA coils, shortens and thickens, and the chromosomes can be seen with a light microscope. The duplicated homologous chromosomes (maternal and paternal) pair up exactly, gene for gene (synapsis), and move together to attach to the nuclear membrane. Crossing-over occurs (see Section 5.1.1 and Figure 5.2). Meiotic spindles (microtubules) and associated proteins form at opposite poles on the cell.
- **Metaphase I:** The nucleus is absent. DNA is at its maximum condensation. The pairs of homologous chromosomes (tetrad) reach and are arranged at the metaphase plate, with the chromosome pair on either side of the plate. Spindle apparatus is fully formed. Centrioles have reached opposite poles. Spindle fibers from each centriole attach to one chromosome of a matching pair. The arrangement of the chromosome pairs with respect to one another at the metaphase plate is completely random. Considering

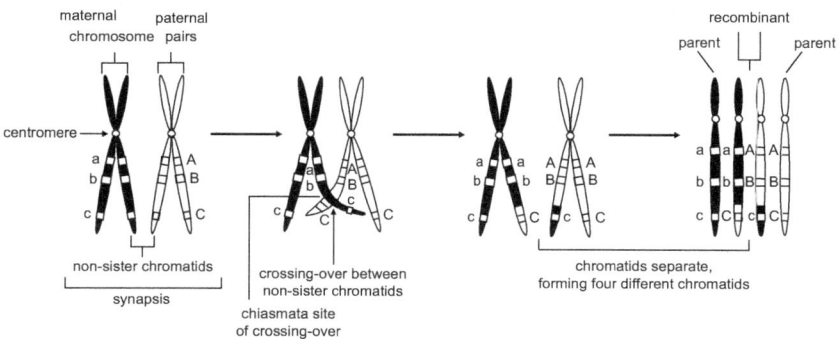

Figure 5.2   Representation of crossing-over, also termed recombination, showing the movement of genes between the chromosomes of the tetrad and the resultant distribution of genes in the final four daughter products. Only one chiasma is represented. The four products are different and comprise two non-recombinant chromosomes with the gene distribution abc and ABC, the same as with the parents, and two recombinant chromosomes, which contain a gene mix of abC and ABc.

the 23 pairs of chromosomes in humans, this means a very considerable number of different arrangements are possible.

- **Anaphase I:** The cell elongates and the first signs of division appear. The two tetrads separate as the chiasmata slip apart (primary disjunction). The chromosomes (sister chromatids) start moving towards the opposite poles (spindle action). The sister chromatids remain connected at the centromere.

Spindles use their contracting action.

- **Telophase I:** A new nuclear envelope forms around the separated chromosomes, which have migrated to opposite poles. Homologous pairs of chromosomes reach either of the two poles. The spindle disappears. Cleavage furrows start to appear.
- **Cytokinesis I:** Complete pinching at the equator produces two cells, each with a complete nucleus and cytoplasm. The daughter cells have a single set of chromosomes.
- **Interphase II:** This is of very short duration in humans but may be extended in other animals. It is often omitted in descriptions or diagrammatic representations of meiosis. *The important point is that there is no replication of DNA during this phase, unlike interphase I.*
- **Meiosis II:** We now have to consider two cells. Comparison of meiosis II (Figure 5.1) with mitosis shows a strong similarity in the steps of both processes. Like mitosis and unlike meiosis I, in meiosis II chromosomes do not attach to the nuclear membrane and cannot form tetrads, thus discouraging crossing-over. The sister chromatids line up in single file in metaphase II on the metaphase plate and are separated at anaphase (secondary disjunction) just as in mitosis (Figure 3.13). The basic difference is that there are pairs of chromosomes (two sister chromatids) to consider in mitosis and only one chromosome (which may be the maternal or paternal one) to break apart per cell in meiosis II. Two cells develop from the one cell in meiosis II, the same as with mitosis. Each cell, however, has one set of

chromosomes (haploid) compared with the two sets of chromosomes (diploid) in each cell in the case of mitosis. Once cytokinesis is complete, there are four sperm cells and one egg cell. In the latter case, the other three cells are small polar bodies that do not develop into eggs (Section 5.2).

5.1.1   Crossing-over in Meiosis

At the prophase I stage, the pair of homologous chromosomes are held together (synapsed) at one or more easily discernable points, each termed chiasma (plural: chiasmata). The structure is a tetrad (Figure 5.2) but often termed a bivalent since the individual sister chromatids cannot be distinguished. When two or more chiasmata are involved, as is often the case, the complexity of gene transfer increases. Humans average just over two chiasmata during meiosis.

It is because of the side-by-side configuration of non-sister chromatids in the prophase/metaphase of meiosis that sections of DNA, and therefore gene information, can be exchanged between homologous chromosomes, and thereby a genetic mix imparted to the final gametes and zygotes. The process is rarely encountered in mitosis because of the linear, and not side-by-side, line up of the chromosomes in mitotic metaphase before separation. Indeed, crossing-over is not desirable in mitosis since absolute integrity is required in DNA duplication. It is, however, a key feature and essential for successful meiosis. Crossing-over is achieved by a breakage and rejoining sequence. It must be carried out very carefully without any net loss or gain of DNA between the chromatids, even as little as a single nucleotide! Failure results in a frameshift in the DNA (Section 4.3.4 in Chapter 4) with attendant problems.

Unequal Crossing-Over

Unequal crossing-over during synapse in meiosis I (Figure 5.3) can lead to a loss or gain of small or large DNA segments and is an important reason for damaging mutations, for example, in Charcot–Marie–Tooth disease.

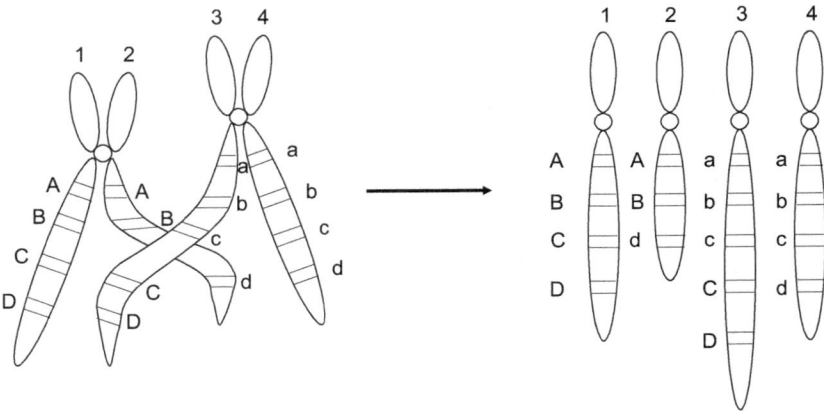

Figure 5.3 Unequal crossing-over in meiosis. The tetrad is non-aligned during synapsis. The single, unequal crossover between chromatids 2 and 3 results in deleted (2) and duplicated (3) chromosomes.

Charcot–Marie–Tooth (CMT) Disease

This disease is caused by a variety of mutations in many different genes. These genes encode different proteins that are concerned with the covering and protection of peripheral nerves. These connect the brain and spinal cord to muscles and to sensory cells. It is the most common neuromuscular disease. With CMT disease there is a gradual wasting of distal limbs and sensory loss from an early age. The *peripheral myelin protein 22 (PMP22)* gene on chromosome 17p encodes the peripheral myelin protein 22, a part of myelin that is located in the peripheral nervous system. CMT type 1A disease arises most commonly from a duplication mutation (Section 4.4.1 in Chapter 4) and leads to an excess of myelin. The duplication of a repeat sequence either side of a 1.5 Mb segment that contains the gene is due to non-allelic homologous recombination during meiosis I (Figure 5.3). There is a concomitant deletion mutation and possession of one copy of this leads to a milder peripheral neuropathy, called hereditary neuropathy, with liability to pressure palsies.

5.1.2   Genetic Linkage

Linkage is the tendency for two different genes on the *same chromosome* to remain together during the separation of homologous chromosomes at meiosis I. If in Figure 5.2 the a and b genes are assumed to be close together on the same chromosome (less than 50 million bp apart), it is unlikely that a recombination event will occur between them. The a and b genes are considered linked and remain together after crossing-over. On the other hand, b and c may be much further apart and it is more likely that they will be separated during the crossing-over, as shown in Figure 5.2. In this case, new allele combinations (abC and ABc) will be present in the gametes but are not shown in the parents. Exchanging portions of adjacent DNA in crossing-over, as well as the random orientation of maternal and paternal homologs at the metaphase plate in meiosis I, ensures that there is very strong diversification of gene content in sex cells, the embryo and the child. Linked genes also have ramifications when inheritance patterns involving more than one characteristic are considered (Section 6.2.5 in Chapter 6).

5.2   GAMETOGENESIS

Meiosis could easily be termed gametogenesis, a process in which gametes are formed. In females, these are eggs, which are formed in the ovaries, and the process is termed oogenesis. In males, these are sperm, which are formed in the testes, and the process is termed spermatogenesis. Representation of the two processes is shown in Figure 5.4. Both mitosis (growth and maturation) and meiosis are involved. We shall also begin to consider the combination of sperm and egg in two animals mating, as well as the disposition of chromosomes and the multifarious genetic consequences for the offspring.

In spermatogenesis, diploid spermatogonia (cells) in male testes mature to become four haploid sperm cells. This takes place in mammals in 30–78 days and in humans over about 74 days. In oogenesis, diploid oogonia (cells) in the ovaries, which are present at birth, mature to become one haploid mature egg (a process occurring in about 30 days). The lower number of viable cells produced in oogenesis is the striking difference from spermatogenesis. With the female, at each division in meiosis I and II, one of the two eggs has more cytoplasm and is larger than the other polar body. The smaller

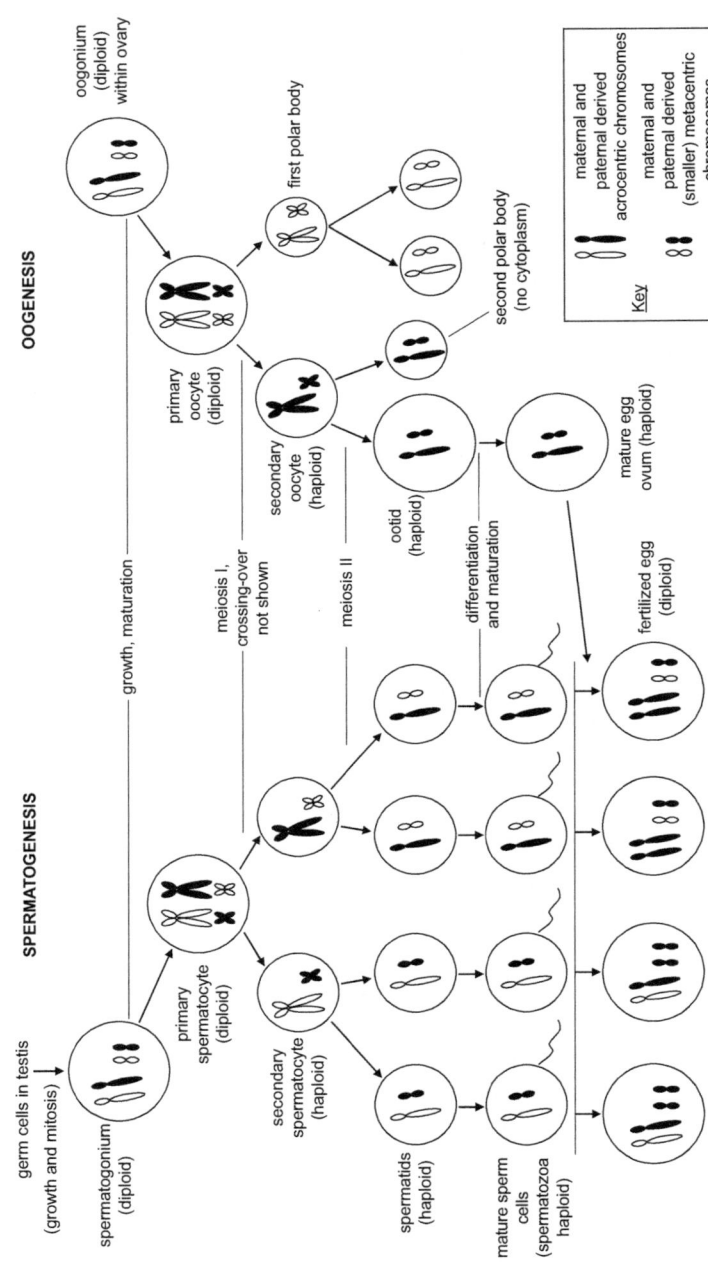

Figure 5.4 The formation of four functional male gametes and one functional female gamete and their associated chromosomal arrangements in animal spermatogenesis and oogenesis. Only two different pairs of chromosomes of the 23 pairs in humans are considered. The longer pair is acrocentric and the shorter pair is metacentric (Section 1.4 in Chapter 1).

cell degenerates, leaving, overall, one large female egg. Any of the four haploid male cells resulting from spermatogenesis can fertilize the single haploid egg to form a single diploid zygote cell, Figure 5.4. The overall process ensures that the total number of chromosomes remains unchanged from parent to offspring. In humans, there are one or two mature eggs awaiting fertilization but as many as ten in mice.

5.2.1   Horses and Mules

It is necessary that, in the meiosis I phase of gametogenesis, there is sufficient matching of the same genes in the same order (synapsis). Crossing-over and the subsequent stages can then occur, ensuring gametes formation and fertility (Section 5.1.1). This is no problem, for example, in the mating of two domestic horses. However, consider the mating of a domestic female horse ($2n = 64$) and a male donkey ($2n = 62$). In addition to the differences in chromosome numbers, only a little over half of the two animal genomes match. Nevertheless, there is apparently sufficient overlap in meiosis I so that mating can occur and a mule may result. Mule and horse/donkey chromosomes are so different that overlap of their chromosomes cannot occur in meiosis I and the mule is sterile. There have been extremely few authentic cases reported of mules giving birth. It is so rare that Romans had a saying, "*Cum mula peperit*" (when a mule foals), for any very unlikely happening. Interestingly, it is different with the domestic horse ($2n = 64$) and the Mongolian wild horse ($2n = 66$). Mating between these can occur and the offspring ($2n = 65$) can also breed.

5.2.2   Errors in Gametogenesis

Mistakes in the formation of gametes (gametogenesis) are the major causes of genetic diseases in humans. It is estimated that 10–25% of all human fertilized eggs contain chromosome abnormalities arising from errors in meiosis and are the major cause of pregnancy failures.

5.2.3 Aneuploidy

Nondisjunction

The most common chromosome abnormality (aneuploidy) is caused by nondisjunction in meiosis. Chromosome homologs may fail to separate correctly during meiosis I, or sister chromatids may fail to separate correctly during meiosis II (Figure 5.5).

Nondisjunction is more common in meiosis I than in II, and both occur as pre-zygotic events. They occur much more frequently than in post-zygotic mitosis, see Figure 3.14. The result is that gametes receive one fewer or one extra copy of a particular chromosome in the fertilized egg. As the resultant embryo develops, the situation persists in every cell in the body. The possible distribution of the sex chromosomes in the event of nondisjunction occurring in meiosis I or II, both in oogenesis and spermatogenesis, is shown in Figure 5.5.

5.2.4 Human Disorders

There are a number of human disorders arising from nondisjunction errors involving the sex chromosomes, but they are rare, occurring in only about 1000 to 2000 live births. Ninety percent of cases are non-inherited, being random events during cell division in early fetal development. They include the following.

Triple X Syndrome 47, XXX

Only affects females; they are fertile, usually tall and have some learning problems. The disorder results from fertilization of an XX-containing egg by an X-containing sperm (Figure 5.5A) or of an X-containing egg by an XX-containing sperm (Figure 5.5B).

Turner Syndrome 45, XO

Only affects females; they are usually short and infertile. It arises in both meiosis I and II, as well as in oogenesis and (usually) spermatogenesis (XO in Figure 5.5A and B). It is the only viable monosomy in humans, affecting 1 in 2500 newborn females but is more frequently found prenatal. See also Section 4.4.6 in Chapter 4.

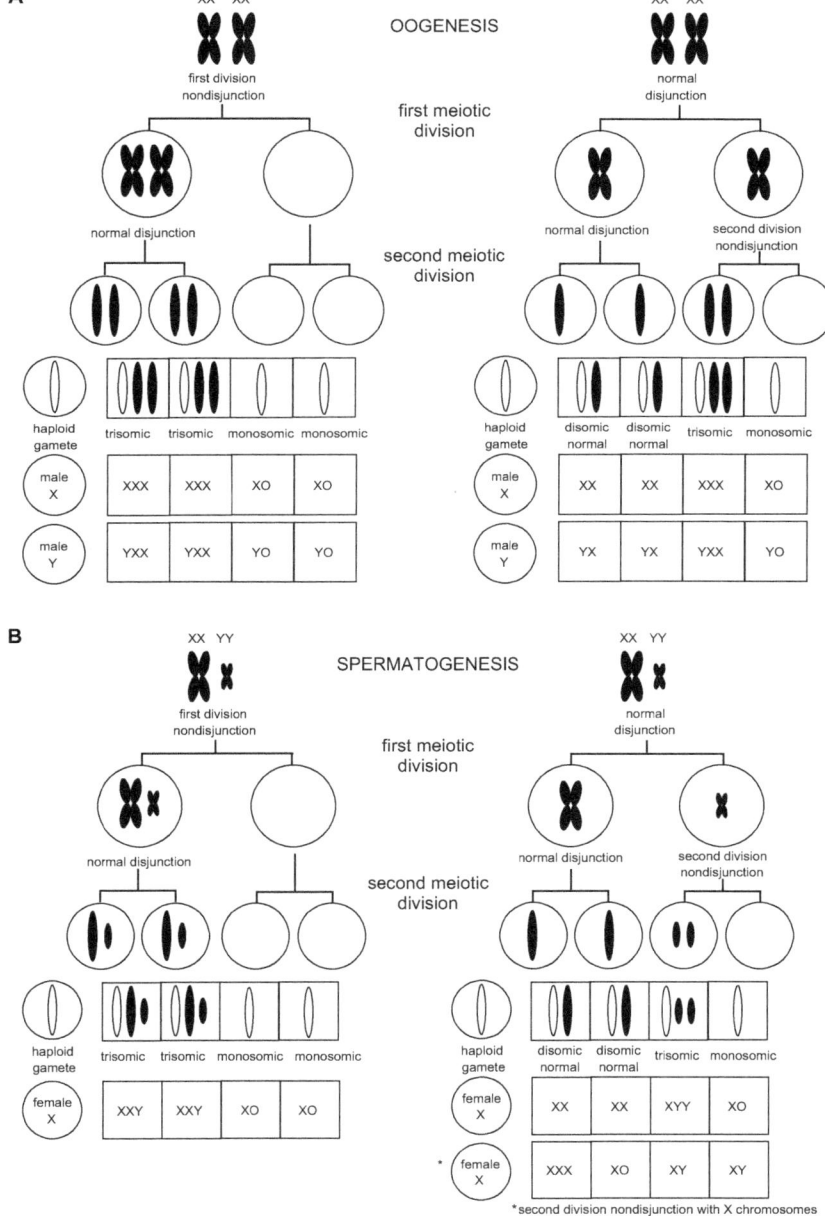

Figure 5.5 Possible outcomes of nondisjunction of sex chromosomes in meiosis during gametogenesis. (A) Oogenesis and (B) spermatogenesis.

Klinefelter Syndrome 47, XXY

Only affects males; they may have small testes and failure of normal sperm production. Many show only mild symptoms. It can only occur in the first meiotic division of the father (Figure 5.5B) but with meiosis I and II in the mother (Figure 5.5A). It is the most common aneuploidy after Down syndrome. About 20% of men with the syndrome have the mosaic version.

Jacob Syndrome 47, XYY

Only affects males; they are fertile, taller than average and have delayed development of motor skills. It can only arise from nondisjunction in meiosis II in spermatogenesis (Figure 5.5B).

All four disorders have a mosaic version with an abnormal chromosome arrangement only present in some of the cells (Section 3.4.4 in Chapter 3).

Trisomy 21

Nondisjunction can also involve non-sex chromosomes. This is shown, for example, in Figure 5.5A, where now the two Xs represent the pair of chromosome 21, and XXX or YXX is the resultant chromosome distribution in trisomy 21. About 95% of people with trisomy 21 (Down syndrome) result from non-disjunction in meiosis, 80% of these being in meiosis I. Most (90%) meiotic errors are of maternal origin in oogenesis. Only about 5% of trisomy originate from nondisjunction during spermatogenesis. Prior to, or at conception, a pair of chromosome 21 in the egg, or occasionally the sperm, fail to separate and trisomy 21, persisting in all cells, results. Humans with trisomy 21 have unusual facial features, some mental problems and a much reduced life expectancy, with sufferers usually dying from heart disease. It occurs in about 1 in 700 live births. In contrast, children with analogous diseases, trisomy 18 (Edward syndrome) and trisomy 13 (Patau syndrome), have multiple congenital defects and die within months or days. The mosaic version of these diseases arising from mitotic disjunction have generally more moderate symptoms but unfortunately account for only 1% of sufferers (Section 3.4.4 in Chapter 3). The remaining 4% of persons with Down syndrome originate from a translocation error (Section 4.4.3 in Chapter 4).

5.2.5 Cattle Disorders

Cattle have 60 chromosomes, 29 pairs of autosomes and one pair of sex chromosomes (XX or XY). All have the centromere at the end of the autosomes and the centromere at the middle of the sex chromosomes. They have aneuploidy disorders that parallel those shown by humans, but the disorders are usually not lethal, although they affect cattle fertility. Some countries and breeding associations require that imported animals or semen show normal karyotypes before calves born there can be registered.

5.3 FERTILIZATION

Fertilization is the fusion of egg and sperm, which triggers the development of the animal. There are a number of steps from the association of sperm and egg to the creation of the fertilized egg.

5.3.1 Pre-sperm Entry into an Egg Cell

Millions of sperm are deposited in the vagina in humans, cattle and other animals, but directly in the uterus in horses, rodents and others (Figure 5.6).

Sperm travel from the vagina, through the cervix, uterus and Fallopian (uterine) tubes, to the fertilization site. This is the ampulla in the middle part of the Fallopian tubes, which contain smooth muscle used to move the egg. Only a few thousand sperm reach the uterine tube and very few sperm get close to the egg at the ampulla. After several hours in the female, physiological alterations in the freshly ejaculated sperm render it hyperactively mobile (termed capacitation) and this improves its ability to penetrate the egg. The egg, which matured in the ovary (secondary oocyte; Figure 5.4), has proceeded down the uterine tube to the ampulla and is arrested in metaphase of meiosis II. It is always the secondary oocyte (Figure 5.4) enclosed in the zona pellucida (Figure 5.7) that is fertilized. Both movement of sperm and the less mobile egg are aided by muscle contractions and

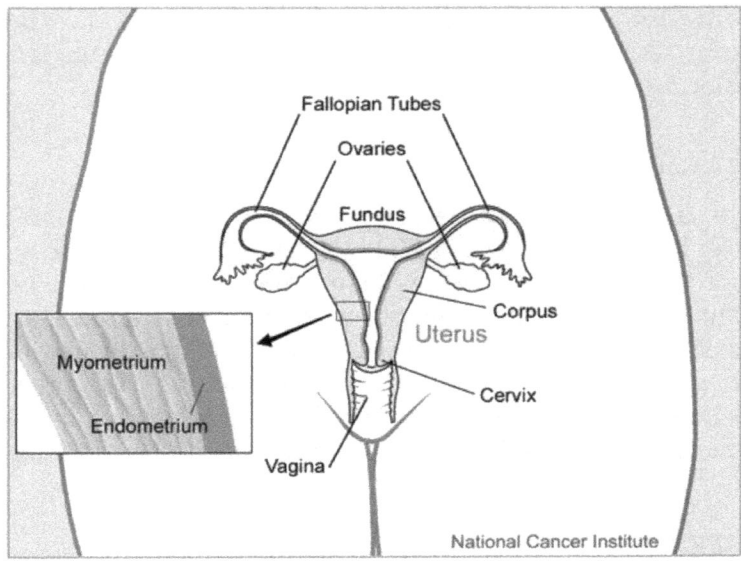

Figure 5.6 The female reproductive organ.
Image courtesy of National Cancer Institute/NIH Medical Arts.

ciliary waving of the tube lining. The egg is activated upon sperm binding. This entails completion of meiosis II and changes in the zona pellucida (Figure 5.7), which discourage more than one sperm fertilizing the egg. Otherwise, polyspermy, which is observed in birds, results in early embryonic death in humans.

5.3.2   Sperm Entry into an Egg Cell

Finally, one sperm enters the egg cytoplasm by a series of steps (Figure 5.7).

These steps entail binding to various barriers in the cell and their subsequent penetration. The entry is aided by the force of the sperm's flagellating tail and its zona-digesting enzymes in the acrosome (Figure 5.8). The binding to the zona pellucida is very species specific, ensuring that, for example, mice sperm cannot fertilize a human egg.

When the sperm reaches the final egg envelope (perivitelline space; Figure 5.7), the egg membrane assimilates the sperm head, allowing the sperm nucleus containing the male nDNA to enter the egg cytoplasm. It is termed the male pronucleus and

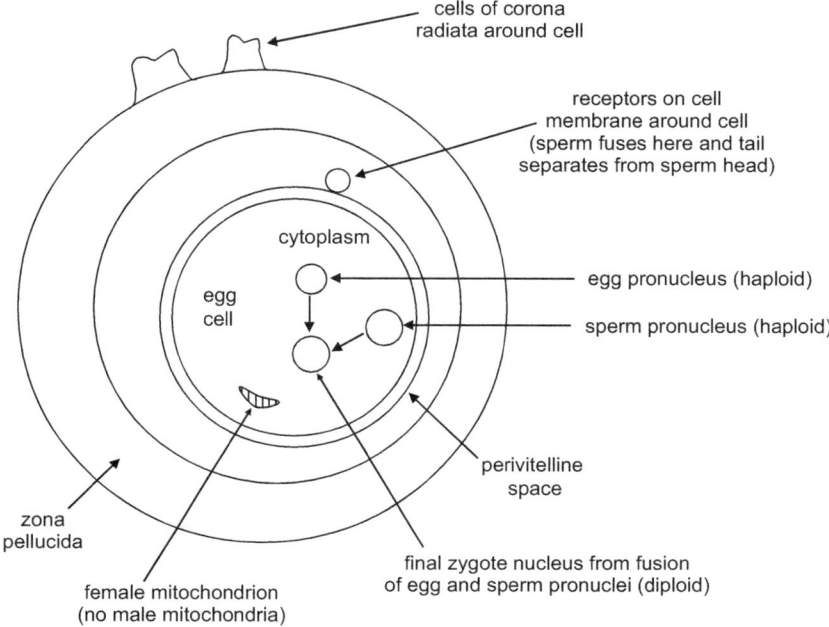

Figure 5.7   The structure of the egg cell and barriers to sperm entry.

is separate from the female pronucleus for a short time (Figure 5.7). The sperm mitochondria (Figure 5.8) either does not penetrate the egg or is degraded. This means that the mitochondrial DNA from the male is, by and large, not used in the egg and the female is the only source of mitochondrial DNA, which has extremely important consequences (see Section 6.4 in Chapter 6).

5.3.3   Early Development of the Human Fetus

The early development of the mammalian embryo for the male and female is similar, with no apparent differences in the sex characteristics. The gender has, however, been set by the X or Y chromosome in the sperm (Section 6.1.1 in Chapter 6). The development process has been studied with humans and also using the fruit fly, nematode worm, African clawed toad, zebra fish, chick and (mainly) the mouse. The zebra fish is a relative newcomer to the scene. It breeds in large numbers and has

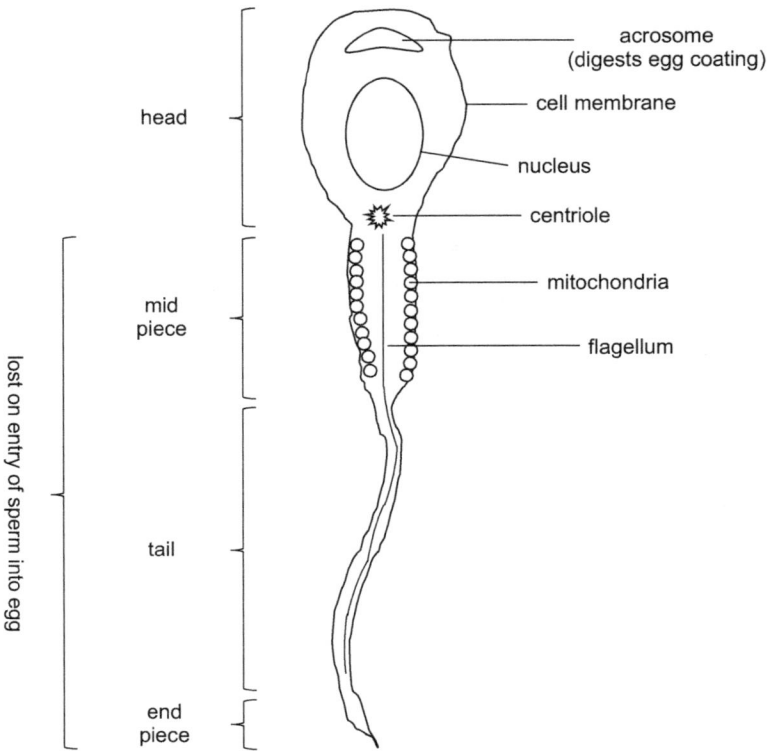

Figure 5.8 The structure of the sperm. In the case of humans, for example, there will be 23 chromosomes in the nucleus in the sperm head.

transparent embryos, thus allowing easy visual observation of cell division. The sperm pronucleus and the egg pronucleus, both with their DNA, approach each other in the center of the cell and fuse within about 36 hours in the case of humans (Figure 5.7). The fertilized egg, termed the zygote, now enters the first stage of mitotic division and cleavage. At the early stages after this it becomes the embryo. When the embryo has reached the uterus, the original one cell has become a blastocyst (Figure 1.3), which attaches to the mucosa lining in the uterus. The blastocyst contains an inner cell mass that develops into the fetus (baby) and an outer wall of cells (trophoblast), which helps form the placenta that nourishes the growing fetus. At this stage, the inner cell mass consists of pluripotent stem cells (see Section 1.1.2 in Chapter 1). A day or two later, the embryo attaches to the uterus. At the end of the eighth week after gestation, the embryo is now termed a fetus, with the major parts of the body beginning to resemble the final animal.

5.3.4   *In Vitro* Fertilization (IVF)

This procedure is complex and used mainly where a woman or man has reproduction problems (*e.g.*, Fallopian tube blockage or poor sperm quality). In one IVF approach, suitable mature eggs from a woman's ovaries, which have been hormone-stimulated to produce more than the usual one egg (hyperovulation), are removed from the ovaries with a fine syringe. If the sperm are impaired, a technique called intracytoplasmatic sperm injection (ICSI) is used to fertilize the egg, whereby a single sperm is microinjected directly into the cytoplasm of the egg in a culture dish under a microscope. After 5 days of incubation to the blastocyst stage, the embryo is transferred back in to the woman's uterus, where it is attached to the mucosa and pregnancy can continue normally. About 20–30% of fertilized eggs result in the birth of children by IVF. Over four million children have been born since the introduction of IVF in 1978 and have become healthy parents themselves.

Preimplantation Genetic Diagnosis and Screening

In recent years, a start has been made, with some successes, to assess cells at the blastocyst stage of IVF. The cells are removed and tested using shotgun sequencing of cell-extracted DNA for a specific genetic condition or for simply confirming normal chromosome numbers. A satisfactory cell can then be selected and transferred to the uterus. The percentage of human eggs with aneuploidy (Section 5.2.3) is 25% with 30-year-old females and as high as 90% with 44-year-old females. The procedure is likely to be helpful for IVF patients who are therefore 38 years or older, as well as those with a history of recurrent miscarriages or inherited genetic diseases.

CHAPTER 6

The Inheritance Patterns of DNA (Chromosomal) Mutations

6.1 FAMILY TIES

So far, we have been concerned mainly with genetic structure and properties of a single animal. Now we consider the mating of animals and, in particular, the consequences of any genetic changes (DNA mutations) for their descendants.

6.1.1 Likely Sex of Offspring

First, the simple case of the likely gender of the infant from two parents is examined (Figure 6.1).

The possible gametes from the male are combined with those from the female. A sperm cell containing an X chromosome that fertilizes an egg containing an X chromosome will, of necessity, produce an XX-containing zygote. This would develop into a female. If, however, the sperm cell contains a Y chromosome, which is as likely, the resultant zygote will contain a male XY combination. It is obvious from this that the male determines the gender of the offspring. So, Henry VIII should not have held his "spouses" responsible for any lack of a male heir! The ratio of male : female births should be, in theory, 1.0. In the US and elsewhere, it is 1.06–1.08 over a large population. This *may* arise

Animal Genetics for Chemists
By Ralph G. Wilkins
© Ralph G. Wilkins 2017
Published by the Royal Society of Chemistry, www.rsc.org

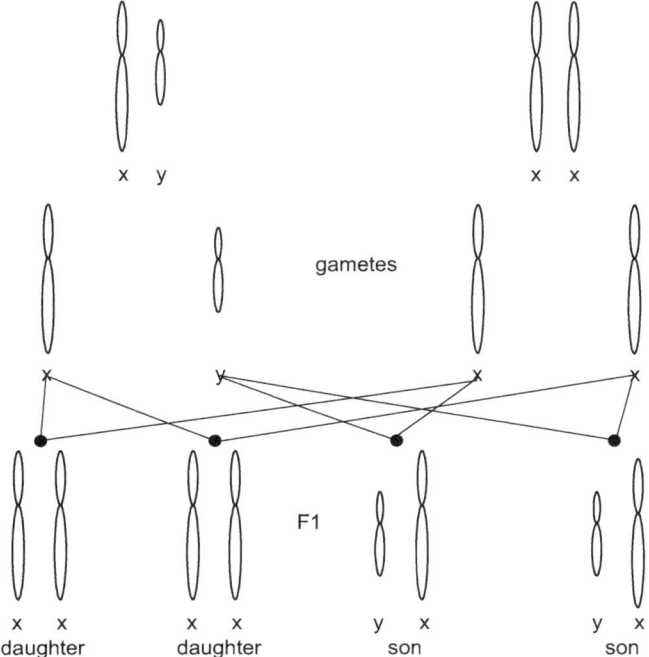

Figure 6.1 The gender, in theory, of the offspring from parents. The four possible crosslinks involving the two female X gametes pairing with either the X or the Y male gamete are shown.

from the sperm carrying a smaller Y chromosome swimming faster than sperm carrying an X chromosome and reaching the egg quicker. An XY combination is then more likely to result.

6.1.2 Family Tree

Some of the symbols used in this chapter are shown in a hypothetical family tree in Figure 6.2.

First-cousin marriages, say between a paternal granddaughter and maternal grandson (shown in Figure 6.2), are illegal in many US states but not in the UK. Bach, Darwin and Poe all married cousins. The dangers from first-cousin marriages, at least on genetic grounds, appear minimal. Surveys show that children from first-cousin parents carry a risk of congenital defects only 2% higher than with offspring of non-related parents. Furthermore, the risk of infant mortality is just 4.4% higher. Similar

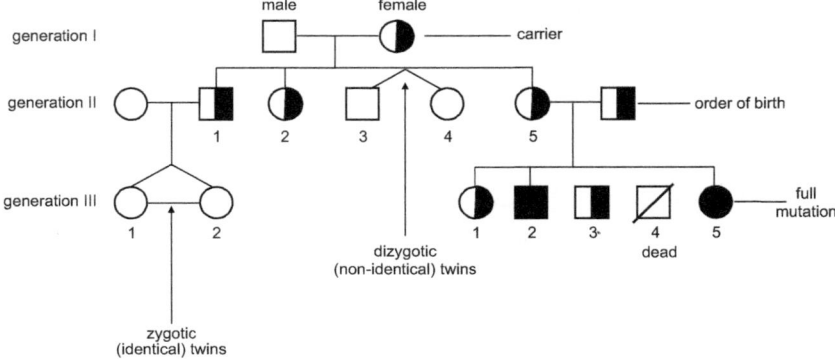

Figure 6.2   Hypothetical family tree. The grandmother is unaffected but is a carrier of an abnormal phenotype (trait; half-circular black symbol). A granddaughter and grandson in generation III show the full trait (solid black circular and square symbols, respectively).

statistics apply to women over 40 who conceive and no one suggests they should not become parents.

Mating that is common in certain populations involving siblings, mother–son and father–daughter (consanguineous) can, however, have more serious consequences. Such inbreeding can bring together two copies of rare, recessive, mutated genes, which previously existed only in one copy in the normal population (and were, therefore, relatively harmless). Certain types of deafness, metabolic disease and skeletal disorders have only shown up in the offspring of such pairings and these disorders were not present in the parents. Any animal only inherits approximately 3% of genetic material from each of its two great-great-great-grandparents. This means that genetic testing should only be used to supplement pedigree and other characteristics in assessing, for example, a horse's racing performance.

6.2   INHERITANCE PATTERNS INVOLVING AUTOSOMES

Now we consider the impact of a faulty gene or genes in either or both parents on their offspring. Here, we are concerned with all but the sex chromosomes, which will be considered later in Section 6.3.

Such a mutant gene may lead to a harmless, perhaps even a desirable, phenotype (a redhead!), but quite often a serious animal disorder results. How and where this mutant gene ends in successive generations and its impact is obviously a very important consideration. Human and animal traits and disorders are used to illustrate the types of inheritance that are possible. The diseases in this chapter can arise from an inherited mutation from an affected parent or both parents, or as a result of a sporadic mutation during the formation of the egg or sperm or the early embryo. In the latter case, there will be no history of the disease in the family. Monogenic diseases arising from mutations in only one gene are rare, but a number feature in this book (Figure 1.13). The genetic repair of these has been addressed over the more prevalent and complex polygenic diseases that arise from mutations involving two or more genes.

6.2.1 Dominant Pedigree

The dominant/recessive relationship of alleles has already been briefly introduced (Section 1.6.3 in Chapter 1). The theoretical inheritance patterns are shown in Figures 6.3A and 6.4A and B, and a possible family tree for an autosomal dominant pedigree is shown in Figure 6.5. In the figures in this chapter the normal gene is indicated with a clear box, while the black box represents a gene mutation that leads to a human or animal disorder (mostly) or, occasionally, an animal coat color or physical appearance. The crosses are termed monohybrid because only one gene and trait is considered.

Characteristics, some of which may be deduced from an examination of the figures, include:

- Males and females are equally affected with frequency and severity, and are equally likely to transmit a condition to sons and daughters. Exceptions to this generalization are sex-limited expression (see Section 7.1.2 in Chapter 7; *e.g.*, ovarian cancer) and sex-influenced expression (see Section 7.1.1 in Chapter 7; *e.g.*, breast cancer).
- The condition appears in successive generations (vertical transmission).

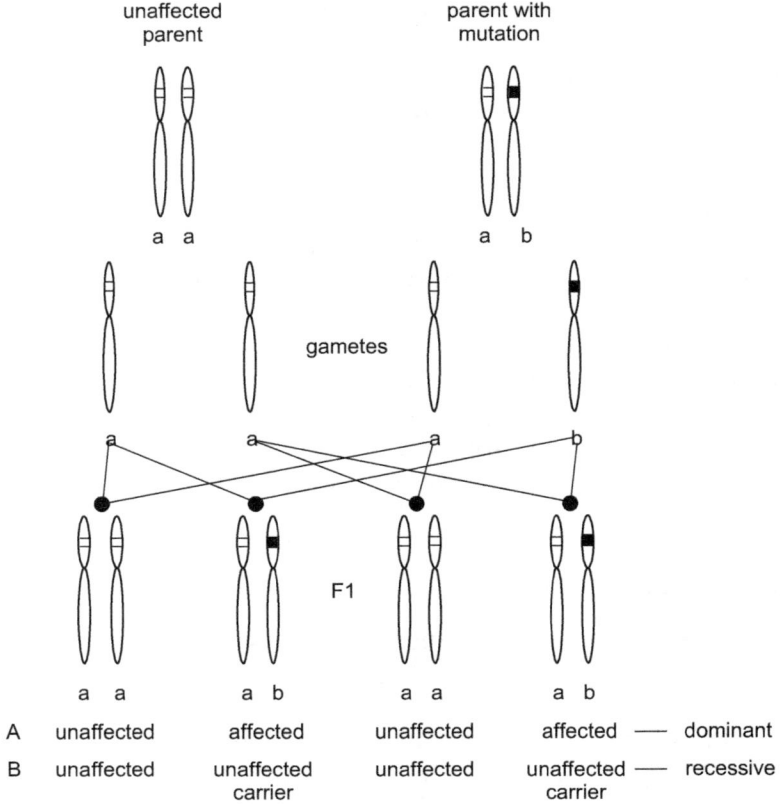

Figure 6.3 One parent affected. (A) Autosomal dominant pedigree. (B) Autosomal recessive pedigree. In this and subsequent figures, the black
allele represents the mutation that may invoke a distinctive physical appearance or a disease.

- A person needs only one copy of the mutation to invoke the
  condition. A person with two mutant alleles is likely to be
  severely affected or even die.
- Each child of an affected and an unaffected parent has a
  50% chance of being affected.
- The chance that two affected persons will have an affected
  child is 75%, comprising a heterozygote (50%) and a
  homozygote (25%).
- The condition may appear at a late age.

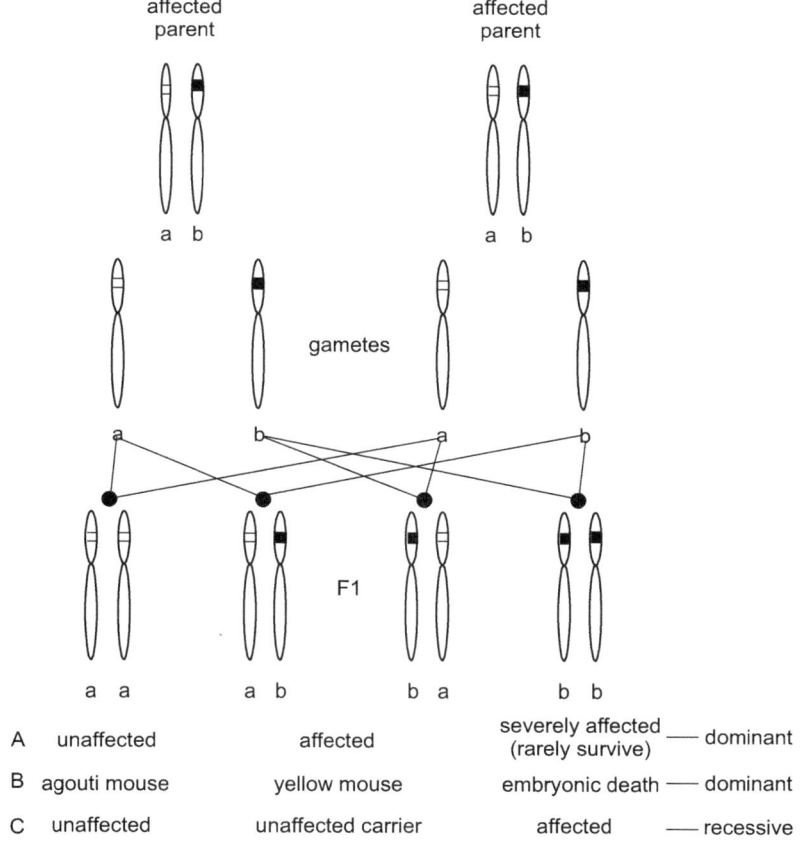

Figure 6.4   Both parents affected. (A) Autosomal dominant pedigree. The same pattern applies to the mating of two lethal yellow mice (B). Normally, one would expect three yellow dominants and one other color. (C) Autosomal recessive pedigree.

- Variable expressivity (Section 7.2.1 in Chapter 7) and reduced penetrance (Section 7.2.4 in Chapter 7) may be exhibited so that some persons show milder symptoms.

6.2.2   Human Disorders: Dominant Traits

Mating of a normal person with an affected heterozygous person for a dominant trait will theoretically lead to normal and affected

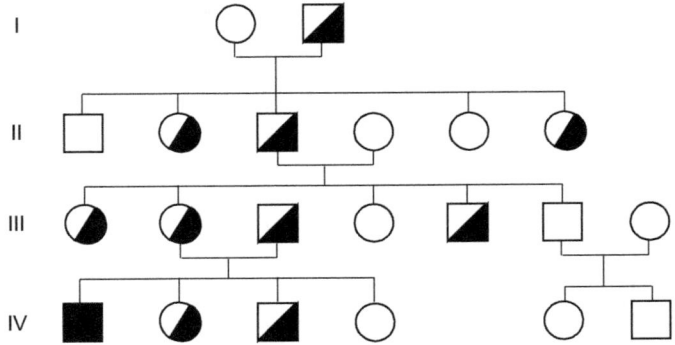

Figure 6.5 Hypothetical autosomal dominant inheritances in several generations, arising from one normal female and one heterozygote
male great-grandparent. The heterozygote (half-black) carries
one copy of the faulty gene and the homozygote carries two
copies of either the normal gene (full-white) or the faulty gene
(full-black).

offspring in a 1 : 1 ratio (Figure 6.3A). Mating of two persons,
each heterozygous, for a dominant trait is undesirable since, in
theory, three of the four offspring are likely to be affected or
seriously affected (Figure 6.4A).

Achondroplasia

The *fibroblast growth factor receptor 3* (*FGFR3*) gene on human
chromosome 4p encodes the FGFR3 protein involved in the
conversion of cartilage to bone and in the maintenance of bone
and brain tissue. The point mutation G380R is involved in 99%
of cases of achondroplasia and probably causes the FGFR3 protein to be overly active and interfere with skeletal development.
This is the most common type of dwarfism in humans and the
characteristic physical appearance is shown in ancient Egyptian
art. Sufferers have short limbs, often a large head and a normal
body. It occurs in about 1 in 15 000 live births and is associated
with older fathers. The autosomal dominant heterozygote always
shows the disorder (complete penetrance; Section 7.2.4 in
Chapter 7) but with a variation in stature (variable expressivity;
Section 7.2.1 in Chapter 7). In about 80% of cases, the disorder
arises as a new mutation and is exclusively inherited from

an older father. Both parents may have normal height. A homozygote with two copies of the mutated gene is either stillborn or dies early. Homologous genes have been identified in mice, zebra fish and other animals.

Familial Hypercholesterolemia

Hypercholesterolemia results from high levels of cholesterol in the blood from both genetic and environmental causes. The excess cholesterol leads to narrowed blood vessels and athero-sclerosis. A small percentage of people with much higher levels of cholesterol and, specifically, low density lipoprotein (LDL—"bad" cholesterol) have an inherited autosomal domin-ant form called familial hypercholesterolemia (FH) from mu-tations of a number of genes. One person in 500 has the heterozygous condition, causing heart problems in middle or later life. One in 1 000 000 have homozygous FH with childhood onset of cardiovascular disease. It is for alleviating the homo-zygous state that mipomersen (Kynamro) has been Food and Drug Administration (FDA) approved (Section 8.5 in Chapter 8).

Other autosomal dominant disorders in humans, mainly inherited, briefly include:

- **Amyotrophic lateral sclerosis (Lou Gehrig disease):** Sufferers have progressive degeneration of the spinal cord and part of the brain, leading to complete paralysis. An unusual char-acteristic is that 90–95% of cases result from a sporadic mutation.
- *BRCA*-**induced breast cancer (Section 7.2.5 in Chapter 7):** This appears to be autosomal dominant because a person inheriting only one mutant allele is at an increased risk of cancer. A second mutant copy is often acquired later in life. Inheritance of two mutant copies of the gene is lethal to an embryo.

These disorders involve more than one gene, whereas Huntington disease, Marfan syndrome, neurofibromatosis type 1, myotonic dystrophy types 1 and 2, and polydactyly, which are all autosomal dominant diseases, are monogenic disorders and described elsewhere.

6.2.3   Animal Disorders

- **Polycystic kidney disease (PKD) in cats and humans:** Multiple fluid-filled sacs (cysts) in the kidneys present at birth, which can grow in size and interfere with kidney function, causing progressive kidney failure and death. It is especially noted in Persian cats and related breeds. The homozygote is probably lethal. There is an autosomal dominant form (type 1) in humans. The protein encoded by the *PKD1* gene on human chromosome 16p has an unknown function.
- **Polydactyly in cats:** This disorder occurs in several animal species but is particularly associated with cats and made famous by Ernest Hemingway. Both the cat and the human version have marked penetrance and variable expressivity (Section 7.2.3 in Chapter 7).

Horses

A number of diseases in horses are autosomal dominant:

- **Equine polysaccharide storage myopathy:** This disease involves excess storage of sugar and results in skeletal muscle disease. Because it is attended by considerable pain, breeding horses that have the disease is not recommended. It is associated with heavy horse breeds, *e.g.*, the Clydesdale, but also occurs in Quarter Horses.
- **Hyperkalemic periodic paralysis:** This is also a muscle disease affecting about 2% of Quarter Horses. Sporadic episodes of muscle contraction and convulsions are observed. The mutation affects the function of the sodium channel in the horse muscle and the condition has been traced to one stallion.
- **Malignant hyperthermia:** This disease has been identified in Quarter Horses, Appaloosas and Paints. The gene mutation causes excessive release of calcium inside muscle cells and can result in fever, irregular heart rhythm and even death when the horse is subjected to anesthesia. Animals with two copies of the mutation rarely survive.

6.2.4   Lethal Alleles in Yellow Mice

Over 100 years ago, experimentalists found that mating very many pairs of yellow mice produced 1063 yellow and

535 non-yellow, *e.g.*, agouti (mottled brown), young mice, with a ratio very close to 2.0. Obviously, the yellow-producing allele (as we now see it) is dominant over the non-yellow allele. Why, then, is the ratio not 3 : 1, as we would normally observe if the dominant/recessive pattern held (see Figure 6.4A, where aa represents agouti and ab, ba and bb represent the dominant yellow allele)? The lower 2 : 1 ratio was explained by supposing that the homozygous yellow mouse (bb) died before birth (Figure 6.4B). This explanation was supported by examining the uteri from pregnant females of such crosses and showing that about one-quarter of offspring died during embryonic development. The yellow is thus a recessive lethal allele, which kills only when the mouse has two such alleles. These ideas were further reinforced by the results from the mating of a yellow mouse (must be ab) with a non-yellow mouse (must be aa). Equal amounts of yellow (ab) and non-yellow (aa) pups would be expected and are observed (compare with Figure 6.3A). The absence of a bb genotype in the pups from this hybrid cross is the reason that dead mice would never have appeared.

Lethal Alleles in Other Animals

Other animals show the 2 : 1 ratio, rather than the normal 3 : 1 ratio, when heterozygote animals are crossed. Examples of heterozygous animals and their mutant phenotypes are:

- Manx cat, with a shortened spine and no tail.
- Bobtail dog, with a short or missing tail.
- Munchkin cat, with shortened front legs.
- Crested duck, with skull deformations in any colored animal.

All show a 25% decrease in litter size compared with animals born from normal (wild type) animals not having the mutation.

Humans

In the animal examples above, only in the homozygous state with two mutant alleles will lethality, or at the very least a very serious condition, be observed. This is not the case, however, in Huntington disease (Section 4.4.1 in Chapter 4), where one copy of the altered allele is sufficient to cause the disorder, which is eventually lethal. Because the onset of the disease is slow, the mutant allele could unknowingly be passed on to progeny. The gene responsible for the disease in humans (*HTT*) also has homologs in mice and puffer fish.

6.2.5   Dihybrid Cross: Guinea Pigs (and Other Animals)

Consider now two characteristics, say black (A) and brown (a) colors and short (B) and long (b) fur. The capital letter represents the dominant allele and the lowercase letter represents the recessive allele. Two different chromosomes might be involved. We can examine both traits of the animal (in this case, a guinea pig) at the same time. Consider the mating of two heterozygous black and short-haired animals, both therefore AaBb. This is a dihybrid cross. During meiosis II, the chromosomes can line up in two different ways before the chromosomes separate (Figure 6.6).

Completion of meiosis yields four types of gametes, namely AB, ab, Ab and aB. Combining these gametes in one parent with an identical set of gametes from the other parent gives the four types of animal shown in Figure 6.7.

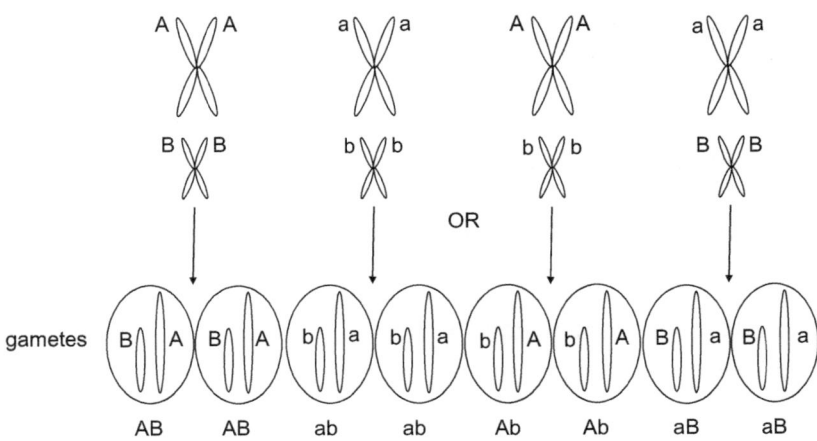

Figure 6.6   Lining up of chromosomes during meiosis. A, a, B and b represent the alleles associated with black, brown, short and long fur, respectively.

		female guinea pig			
	gametes	AB	Ab	aB	ab
male guinea pig	AB	AABB black, short	AABb black, short	AaBB black, short	AaBb black, short
	Ab	AABb black, short	AAbb black, long	AaBb black, short	Aabb black, long
	aB	AaBB black, short	AaBb black, short	aaBB brown, short	aaBb brown, short
	ab	AaBb, black, short	Aabb black, long	aaBb brown, short	aabb brown, long

thus, there are: nine black, short   three black, long   three brown, short   one brown, long

Figure 6.7 A Punnett square for mating of two heterozygous black and short-haired guinea pigs using gametes generated in Figure 6.6. One or two capital letters in the set of four will ensure a black (A) or short-haired (B) animal. The increased complexity of the crosses requires the somewhat tidier Punnett square treatment rather than the previous depictions.

Table 6.1 Two characteristics of various animals.

Animal	Dominant trait	Recessive trait
Cat	Brown coat, short tail.	White coat, long tail.
Cattle	Black coat, no horns (polled).	Red coat, horns.
Human	Brown eyes, black hair.	Blue eyes, red hair.

There are nine black and short-haired animals, three black and long-haired, three brown and short-haired, and one brown and long-haired animal resulting. Again, this ratio would not be likely seen even if the animals produced 16 guinea pig pups! Even so, one will probably see more of the black, short-haired (with the dominant alleles) animals than the brown, long-haired (with the recessive alleles) animals. The $9:3:3:1$ ratio can be checked by considering each trait singly. The black-to-brown ratio is $12:4$, *i.e.*, $3:1$, as in the monohybrid cross. Similarly, the short-to-long-haired ratio is also $12:4$. The $9:3:3:1$ ratio is frequently observed *over a large population* with a variety of animals and characteristics (Table 6.1).

In this treatment, it is assumed that there is independent assortment of the genes, *i.e.*, they are not linked (Section 5.1.2 in Chapter 5), otherwise the alleles will not segregate independently during gametes formation and the phenotype ratio will deviate from these numbers.

Although one allele will often completely dominate the effect of the partner allele of a gene, giving the dominance/recessive pattern, variations of this relationship are observed.

6.2.6   Codominant Pedigree

Sometimes, neither allele is dominant and both alleles are expressed when together in the heterozygous condition. This is in contrast to the dominant/recessive situation, where one allele completely bleaches the other in the heterozygote.

6.2.7   ABO Blood Group System

The first discovered ABO blood group system used by humans is an often-quoted example of codominance. The blood group trait is determined by a single gene, *ABO*, on chromosome 9q. There are three alleles of this gene, A, B and O. A and B are codominant; O is recessive. A child must have two O alleles to have the O blood group. Parents with blood type A (genotype AO) and blood type B (genotype BO) can have, in theory, four children with all four different blood groups, namely AB, A, B and O, with equal likelihood (Figure 6.8). Blood grouping is an early example of precision medicine accounting for individual's variability.

6.2.8   Animal Traits: Roan Cow

The codominant behavior is also well illustrated by the roan coat color shown by many animals. Roan is an even mixture of white (or gray) and another colored hair. Roan in shorthorn cattle is controlled by the *mast-cell growth factor* (*Mgf*) gene on bovine chromosome 5, which is involved in pigment production. When the shorthorn has two copies of a functional Mgf allele, it will be fully pigmented and red (aa). When both alleles have a mutation, the animal is white (bb). Breeding the red with a white animal will always give a heterozygous roan calf (ab) with one copy of a functional *Mgf* allele and one copy of a mutant *Mgf* allele, which means the calf will show both colors and have a red coat with white blotches. When two roan animals are bred, the phenotype ratio of the calves is 1 (red):1 (white):2 (roan), as shown in Figures 6.9A and 6.10.

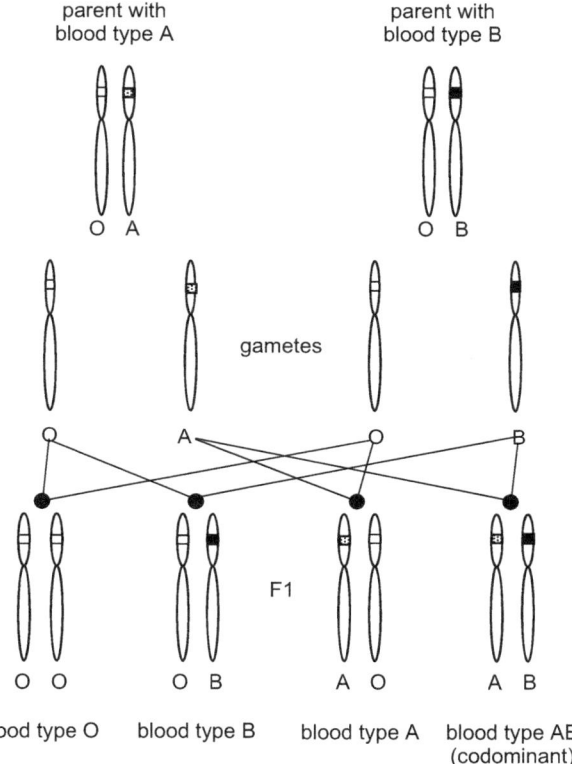

Figure 6.8 Blood types of children from parents with blood types A and B.

Some other examples are shown in Table 6.2. Both colors of the parents are expressed in the offspring.

6.2.9 Incomplete Dominant Pedigree

A third type of relationship between the allelic pair of a gene is incomplete dominance, also known as semi- or partial dominance. Here, the phenotype of the heterozygote is intermediate between the phenotypes of the two homozygotes. It is a new phenotype, unlike codominance, where the heterozygote shows a mixture of the two phenotypes and not a blend, as with incomplete dominance. The concept can be illustrated in the color of certain breeds of horses.

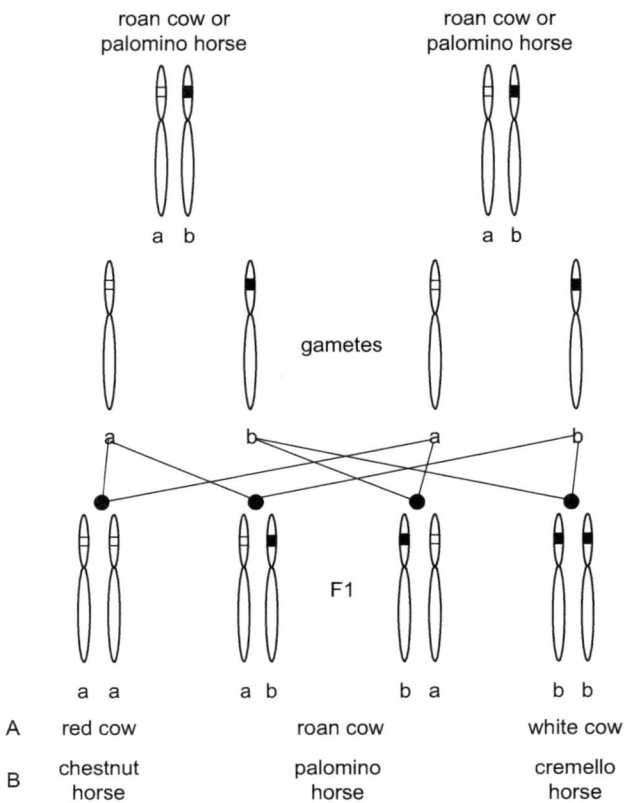

Figure 6.9 (A) Codominant red- and white-producing genes in cows and the formation of a roan cow. (B) Incomplete dominant cremello- and chestnut-producing genes in horses in the formation of the palomino horse.

Animal and Human Characteristics

The cream dilution gene (C locus) in animals consist of a and b alleles. The cremello (b) allele is said to show incomplete domin-ance and dilutes the red in a chestnut horse (aa) to a yellow color with a single dose, giving the palomino (ab), and dilutes further the red to pale cream (in fact, a near white color) in a double dose, giving the cremello (bb) shown in Figure 6.10. Mating a palomino male (ab) and female (ab), you should get about half palominos, with the remainder being chestnuts (aa) and cremellos (bb), *i.e.*, a ratio of 1 : 2 : 1 for chestnut : palomino : cremello foals (Figure 6.9B).

pure red   roan   pure white

chestnut   palomino   cremello

Figure 6.10   (Top) Red, roan and white cows. (Bottom) Chestnut, palomino and cremello horses. Note that the codominant pattern of the roan cow shows both red and white colors, while the incomplete dominant pattern of the palomino is a uniform blend of chestnut and cremello colors, *i.e.*, a new phenotype.

Table 6.2   Examples of codominance in animal colors.

Parents	Heterozygote animal
Black and white (or grey) horses.	Blue roan horse with greyish or bluish body and black extremities.
White and grey horses.	White horse with grey spots on rump and loins (Appaloosa).
Brown and black cats.	Brown cat with black spots or stripes, or black cat with brown spots or stripes (tabby).
White and black chickens.	Chicken with white and black feathers.

Table 6.3   Examples of incomplete dominance in animal and human characteristics.

Parents	Heterozygote animal
Rex rabbit (short fur) and Angora rabbit (long fur).	Rabbit with short fur but not as short as the Rex rabbit.
Black and white mice.	Gray mouse.
Curly and straight hair in humans.	Wavy hair.
Large and small nose in humans.	Medium-sized nose.

A ratio of 1 : 3 for chestnut : cremello foals would have resulted if the cremello were a completely dominant allele. Some other examples are shown in Table 6.3.

6.2.10 Recessive Pedigree

Some of the characteristics of autosomal recessive inheritance can be deduced from Figures 6.3B and 6.4C. These figures apply to all the diseases and traits in this section.

- Males and females are equally affected.
- Unaffected parents, but with one or two heterozygotes (carriers), often have affected children. This is different from the dominant case, where unaffected parents have unaffected children.
- The disorder occurs only in one generation, often involving several children.
- A person with only one copy of the mutant gene is termed a carrier and is unlikely to experience the disorder. This is a happier situation than the dominant type, when unpleasant traits and disorders are expressed with the heterozygote. In the carrier, enough functional protein may still be synthesized, or the alteration of the protein may not be significant so that the normal phenotype and function is unimpaired. A mutant recessive gene usually only shows its effect when an animal has two copies of it. This may lead to a radically changed product or no product at all and, therefore, a severely changed phenotype, sometimes with disastrous consequences.
- Each child of a phenotypically healthy carrier and an unaffected person has a 50% chance of being a carrier.
- The chance that two unaffected (carrier) persons will have an affected child is 25% and a great majority of affected persons arise from this combination.
- The condition may appear in infancy or childhood.
- Variable expressivity and/or reduced penetrance is less common than with the dominant case.
- The disorder may be more common in ethnic groups and of higher frequency in children of consanguineous parents, *e.g.*, first cousins.

6.2.11 Human Disorders: Recessive Traits

There are a large number of disorders associated with autosomal recessive inheritance. In general, they arise from mutations

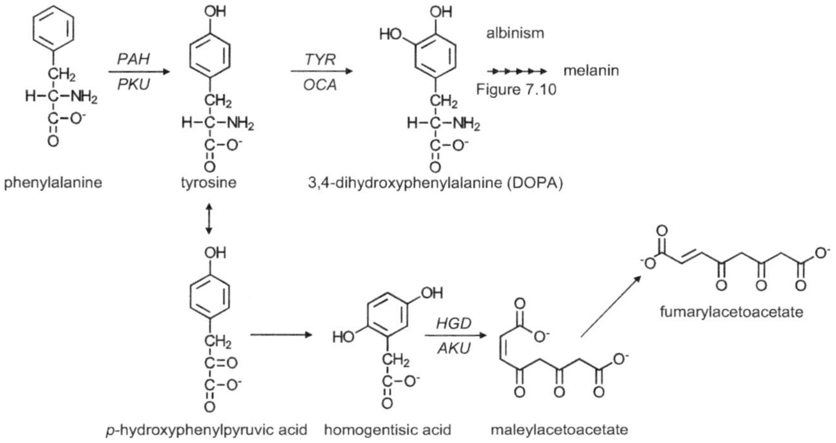

Figure 6.11 Metabolism of phenylalanine and tyrosine, and the location of three genes and diseases associated with mutations of these genes.

leading to faulty enzyme defects. This is in contrast to dominant inheritance, where mutations are more likely to result in ineffective non-enzymes or structural proteins. The first three diseases described in the following pages all result from the modification of genes involved in three steps during the metabolism of phenylalanine (Figure 6.11).

Metabolism of Phenylalanine and Tyrosine

Phenylketonuria (PKU). This is associated with the *phenylalanine hydroxylase* (*PAH*) gene on chromosome 12q, which encodes the liver enzyme phenylalanine hydroxylase. This catalyzes the first step in a series of reactions that convert phenylalanine to compounds used in cellular respiration, as well as the synthesis of the important pigment melanin, Figure 6.11. More than 500 mutations in the *PAH* gene have been linked with phenylketonuria. Most are as the result of a single amino acid change in the enzyme, with R408W the most common mutation in Europe and probably the world. One prevalent mutation designated IVS12 + 1G > A involves a G to A nucleotide change, which, during alternative splicing (in transcription; Section 3.1.4 in Chapter 3), results in the undesirable excising of exon number 12 of 13 exons. The resulting mRNA contains a premature stop codon. The result is a truncated PAH protein that is 52 amino acids shorter than normal, with almost zero activity and effectiveness. As a consequence, phenylalanine can build up to toxic levels in the blood and brain, causing serious health problems. It was the first identified neurogenic disease. There is a wide variation in symptoms from severe classic PKU (R408W and IVS12 + 1G > A mutations) to mild forms (Y414C mutation), all involving nervous system damage and mental retardation. About 1 in 10 000 newborns have the disease. Very early screening and diet control can prevent the disease from developing. Figures 6.3B and 6.4C show the condition of children resulting from one carrier parent or two carrier parents.

PKU and Population Genetics. Examination of the occurrence in countries of specific gene mutations can throw light on the relationships between different populations. Since PKU is one of the most frequent genetic disorders in Europe and is monogenic, it is useful for studying the frequency of specific mutations and their appearance in different countries and to speculate on their origins. In Denmark, North Germany and the British Isles, IVS12 + 1G > A is the most frequent mutation. The mutation may have been brought to the British Isles by Anglo–Saxon immigrants in the first millennium AD. The high frequency of this mutation in North Italy may have resulted from migration from North Germany around the same time. In contrast, the relative rareness of this mutation in the Irish population bears witness to the forced English subjugation! Frequent PKU mutations in Eastern Norway, such as F299C, do, however, appear in Ireland, pointing to Viking migration between the seventh and ninth centuries. The high frequency of the Y414C mutation throughout Scandinavia may explain its appearance in the British Isles, Germany and even as far south as Sicily through Viking and Norman migration. It is speculated that the R408W mutation originated in an ancient Eastern population and spread widely westwards to the Mediterranean and adjacent areas. The Mediterranean countries where migration was common have a broad range of mutation types.

Alkaptonuria (AKU). This is associated with the *homogentisate 1,2-dioxygenase* (*HGD*) gene on chromosome 3q, which encodes the enzyme homogentisate oxidase. This converts homogentisic acid to maleylacetoacetate and thereby helps to break down phenylalanine and tyrosine (Figure 6.11).

More than 65 mutations in *HGD* have been identified. They are mostly one amino acid changes, a common one being M368V. The mutations impair the enzyme action and homogentisic acid builds up, causing the alkaptonuria condition. This disease is characterized by a dark brown discoloration of the skin and eyes, and there is progressive damage to the joints (arthritis and spine) that is sometimes linked in later life to very severe arthritis. The acid in the urine turns dark on exposure to air, producing black-colored urine! It is rare; there are only about 1 in 250 000 worldwide. It was the first metabolic disease studied. Carriers do not typically show signs of the condition in this recessive pattern.

Oculocutaneous Albinism (OCA). There are four types of OCA, each resulting from mutations in single genes and all inherited with an autosomal recessive pattern. About 1 in 20 000 are born with OCA and 1 in 70 are carriers. Type 1 or OCA1 is the most common and is associated with the *tyrosinase* (*TYR*) gene on chromosome 11q, which encodes a copper-containing enzyme called tyrosinase. This catalyzes the conversion of tyrosine to dihydroxyphenylalanine in the first of many steps in the production of melanin (Figure 6.11 and Figure 7.10). There are more than 100 mutations in the gene, which are characterized by white hair, very pale skin and light-colored irides. The homozygous state results in albinism regardless of how many pigment genes are present, *i.e.*, it is an epistasis gene (see Section 7.3.6 in Chapter 7). The defective gene is also a good example of pleiotropy (see Section 7.3.1 in Chapter 7) in that it can produce multiple phenotypes, namely variations in hair, skin and eye color. OCA type 2 (OCA2) is associated with the *oculocutaneous albinism II* (*OCA2*) gene on chromosome 15q, which encodes the P protein, essential for normal pigmentation and is likely involved in melanin production. Up to 100 mutations have been identified with persons with OCA2, which is the most common form of OCA in persons of sub-Saharan heritage. They have similar characteristics to those with OCA1 but are less severe, *i.e.*, cream skin or hair color. The most common mutation is a large deletion (2.7 kb) in the gene. Certain changes in the *MC1R* gene (Section 7.3.9 in Chapter 7) modify the appearance of people with OCA2. People with genetic modifications in both the *OCA2* and *MC1R* genes have many of the features of OCA2 but typically red rather than lighter-colored hair. The P protein generated by the *OCA2* gene plays a major role in the amount of brown pigment (melanin) deposited in the iris of the eye. Production of large amounts of melanin leads to brown eyes. The most frequent color worldwide. Several common variations (polymorphisms) in the *OCA2* gene reduce the amount of P protein produced, meaning less melanin and lighter-colored (blue) eyes. In addition, an SNP in intron 86 of the nearby *HERC* gene controls the expression of the *OCA2* gene as required. These two genes are, therefore, important in establishing the brown and blue eyes of people (Section 1.6.3 in Chapter 1). There is a temperature-sensitive form of the gene in humans analogous to those in animals, leading to OCA-TS (Section 7.3.9 in Chapter 7). Many other human disorders showing autosomal recessive behavior include the following.

Hemochromatosis Type 1

Hemochromatosis type 1 is associated with the *hemochromatosis* (*HFE*) gene on chromosome 6p, which encodes a protein that, with other proteins, detects and helps tightly regulate iron absorption from the diet and iron release from body storage sites. It is an unusual disorder and controlled by one gene. More than 20 mutations in the *HFE* gene can cause hemochromatosis type 1. It is the most common of four types, developing after the age of about 40 years. Most cases of hereditary hemochromatosis among Caucasians are caused by C282Y and H63D mutations. Both these mutations affect how the protein folds and limit its ability to reach the cell and help with iron regulation. Excessive absorption and storage of iron result, which lead to damage and impairment of many organs, for example, the liver and pancreas. Bloodletting or treatments with chelating (metal-binding) agents may alleviate the condition. It is one of the most common disorders in the US, affecting 1 000 000 people. Types 1, 2 and 3 are all autosomal recessive. Type 4 differs in having an autosomal dominant inheritance pattern (Section 6.2.1). Interestingly, not everyone with two copies of the mutated *HFE* gene contract hepatic iron overload.

Sickle Cell Disease (SCD)

This is associated with the *hemoglobin, beta* (*HBB*) gene on human chromosome 11p, which encodes the beta-globin protein. Hemoglobin consists of two subunits of beta-globin and two subunits of alpha-globin. Each of the four subunits contains an iron heme, which binds oxygen in red blood cells and transports it from the lung to the cells in the animal body. A famous point mutation in the *HBB* gene, E6V (GAG, Glu $\rightarrow$ GTG, Val), was the first characterized mutation associated with a human genetic disease. It leads to an abnormal version of beta-globin called hemoglobin S or HbS, which may replace a beta-globin subunit in hemoglobin. The vast majority of people have two copies of the normal gene, *i.e.*, they have the HbA/HbA genotype. If they have two copies of the mutant gene, *i.e.*, are HbS/HbS, they have sickle cell anemia with sickle cell blood cells. In these, the normal donut-shaped red blood cells are bent into sickle-shaped cells, which die prematurely, leading to a shortage of red blood cells (anemia) and can also cause pain and organ damage by blocking small blood vessels. Someone with only one copy of the mutant gene (HbA/HbS) has the sickle cell trait but experiences no anemia. The normal red blood cells containing HbA are sufficient in amount to carry out the required tasks. This heterozygous person has some resistance to malaria (pleiotropy; Section 7.3.1 in Chapter 7), which explains, in part, the high frequency of the mutant gene in malaria-ridden areas. It appears that the abnormal sickle blood cells are not a hospitable environment for the parasite responsible for malaria. SCD mainly affects people of African ancestry. There are hundreds of mutant versions of the *HBB* gene, leading to a group of inherited SCDs with a wide range of symptoms, of which sickle cell anemia is the most common and severe. No satisfactory treatment is yet available, although gene therapy is a viable possibility and is being actively explored; see Section 8.4.10 in Chapter 8 for more. Tay–Sachs disease, cystic fibrosis, Friedreich ataxia and familial lipoprotein lipase deficiency are other autosomal recessive diseases described elsewhere (see also Figure 1.13).

6.2.12   Animal Characteristics

Cheetah Coat Patterns

The common (spotted) cheetah has spots, whereas the king cheetah has spots that have merged into large stripes or blotches. The distinctive coat patterns are associated with the *transmembrane-aminopeptidase Q* (*Taqpep*) gene, which encodes a membrane-bound metalloprotease. It was the first gene found to influence the coat patterns in mammals and probably works by determining the level of another gene that is involved in determining the shades of hair cells.

A gene mutation converts spots into stripes on cheetahs. The wild-type gene (A) in the common cheetah is dominant over the recessive, mutated gene (a) in the king cheetah. Therefore, two copies of the mutated *Taqpep* gene (aa) are required for a king cheetah phenotype, while one copy (aA) leads to a common cheetah. The family tree of captured cheetahs from South Africa is available. A portion of this tree is shown in Figure 6.12.

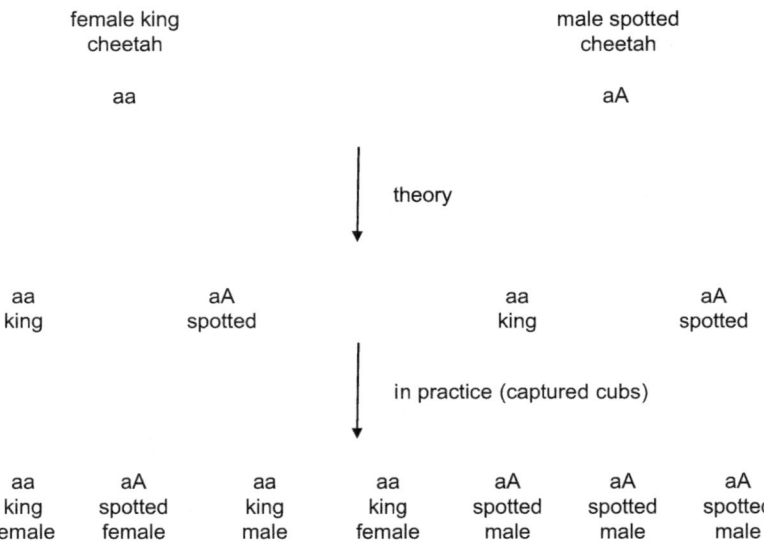

Figure 6.12 Types and genders of cubs resulting from the mating of a common cheetah and a king cheetah in theory and in practice.

It involves the mating of a heterozygous common cheetah (Aa) with a homozygous king cheetah (aa). The 1 : 1 ratio expected in theory for the common-to-king cheetah cubs (Figure 6.12) and male-to-female is almost exactly observed (the actual ratio is 4 : 3). The *Taqpep* gene pattern is set very early in embryo development, and therefore does not change as the animal ages. A different mutation in the same gene is responsible for transforming the most common domestic tabby cat (the Mackerel) with vertical stripes on the body into the classic tabby, which has blotched dark swirls like the king cheetah. Genetically, the classic tabby cat corresponds to the king cheetah and has the recessive gene.

Horse Disorders

Only a few genetic disorders have been found in stock horses. Two of these are inherited as autosomal recessive disorders, as follows:

- Hereditary equine regional dermal asthenia (HERDA) is a skin disease. About 40% of Quarter Horses are carriers of the gene. The inheritance of HERDA conforms to the usual recessive pattern (Figures 6.3B and 6.4C).
- Glycogen branching enzyme deficiency (GBED) causes a late-term abortion or weak foals. About 80% of Quarter Horses are carriers. The disease has been genetically traced to a specific Quarter Horse.

6.3   INHERITANCE PATTERNS INVOLVING THE X CHROMOSOMES

We now have to consider the sex of the parents. The relevant gene is located on the X chromosome of the male or female and, again, as with the autosomal cases, the gene can act either in a dominant or recessive manner, although the distinction is less clear than with their autosomal counterparts. In fact, the two modes are sometimes discussed collectively. Over 100 principle X-linked inherited human disorders have been identified. Most are classified as recessive, a much smaller number as dominant, of which a few are lethal.

6.3.1 X-linked Dominant Pedigree

The inheritance patterns for mating an affected father or mother with a normal partner are shown in Figures 6.13A and 6.14A, respectively. A hypothetical family tree for this inheritance is shown in Figure 6.15.

Examination of these figures reveals the important characteristics of this condition. The X-linked recessive inheritance pattern (Section 6.3.4) has very similar characteristics.

- Females with two copies of the mutant gene are very rare. Both heterozygous females and hemizygous males, each with one copy of the mutant gene, can show the disorder characteristics.

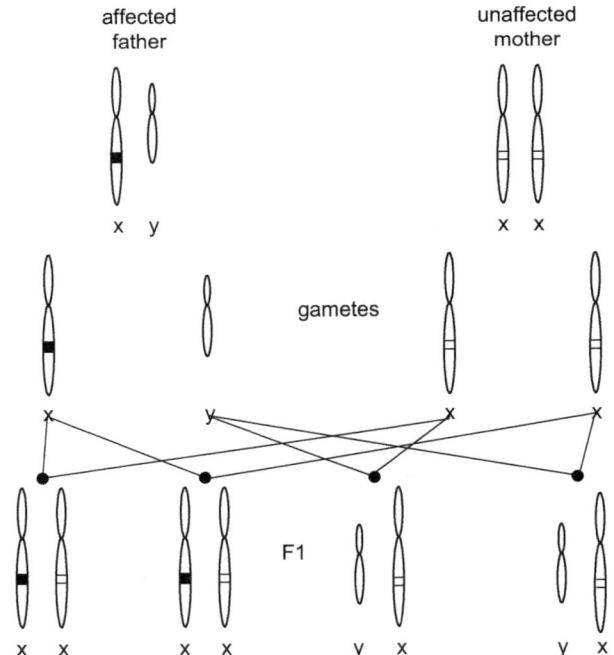

A	affected daughter	affected daughter	unaffected son	unaffected son dominant
B	carrier daughter	carrier daughter	unaffected son	unaffected son recessive

Figure 6.13 Children resulting, in theory, from an affected hemizygous father and a normal mother for (A) an X-linked dominant pattern and (B) an X-linked recessive pattern.

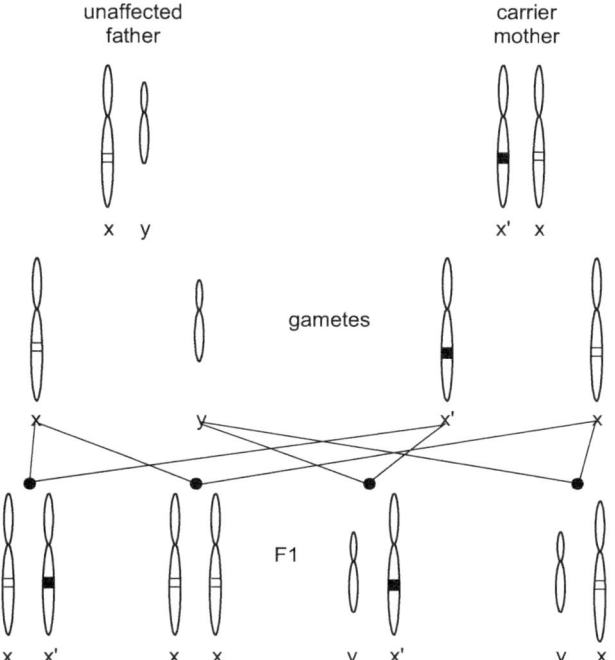

Figure 6.14 Children resulting, in theory, from a normal father and an affected hemizygous mother for (A) an X-linked dominant pattern and (B) an X-linked recessive pattern.

- Both son and daughter offspring of an affected mother and unaffected partner have a 50% risk of inheriting the phenotype (Figure 6.14A). This is a similar pattern to that seen in an autosomal dominant inheritance (Figure 6.3A).
- Affected males with normal partners have no affected sons but always affected daughters (Figures 6.13A and 6.15). The lack of male-to-male inheritance differs from autosomal-based inheritance.
- The effect is shown more often, sometimes exclusively, in the female. This is because the female can inherit one affected X but has, in addition, one normal X chromosome to offset the mutant gene. Affected females have milder but

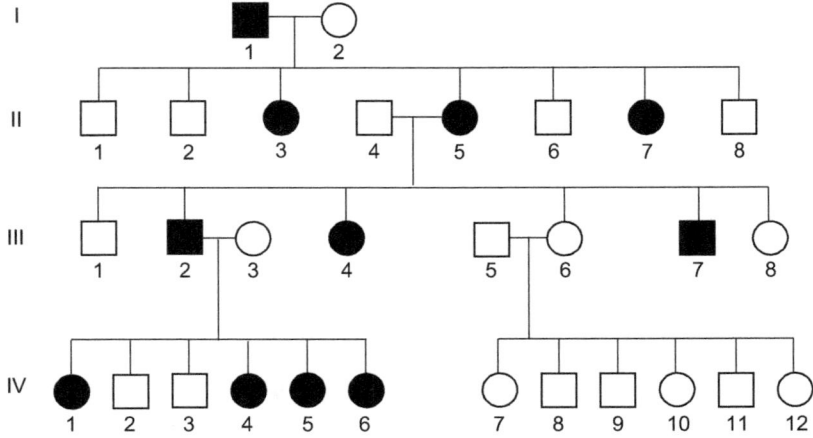

Figure 6.15 Hypothetical family tree originating from an affected male and an unaffected female (great-grandparents). Black squares and circles represent a dominant-controlled male and female, respectively, each with one copy of a mutant gene.

variable expressivity (Section 7.2.1 in Chapter 7) of the trait. Males, in contrast, have only one copy of an X chromosome (hemizygous), and in males the disease is more severe and they often do not survive.

6.3.2   X-linked Dominant Human Disorders

Thankfully, there are relatively few common human disorders associated with an X-linked dominant trait. They include the following.

Rett Syndrome

This involves the *methyl-CpG-binding protein 2* (*MECP2*) gene on human chromosome Xq, which encodes a protein (MECP2) that is critical for normal functioning of the brain, in which it is very abundant, although with an uncertain role. The protein binds to methylated DNA and functions as a master regulator of gene transcription by switching it off (see Section 7.4 in Chapter 7). The vast majority of sufferers with Rett syndrome have a mutation in this gene. More than 300 mutations have been identified. These include single base-pair changes and insertions or deletions, resulting in altering either the structure or the amounts of protein. A mutation may prevent it from binding to methylated DNA, and thus gene expression is disrupted. Rett syndrome is a brain disorder found almost exclusively in girls (1 in 8500), partly because males with the mutation often die in infancy. There is normal early development (up to 18 months) but, thereafter, considerable brain disorders result, causing problems with communication and coordination. Almost all people with classic Rett syndrome have no history of the disorder in the family since it arises from new mutations in the *MECP* gene. The syndrome has been an effective vehicle for studying the role of epigenetics in human disease using the mouse homolog of the human *MECP* gene.

A cloned strain of mice has been created, in which the *MECP2* gene has been inactivated, and therefore MECP protein production terminated. Females developed Rett syndrome and males died. This would be expected from the experience with humans. What was unexpected, however, was that, on removing the inactivation, protein reappeared and mice near death recovered. This reversibility for a neurological disorder was not anticipated and gives hope that curing this and other developmental disorders is at least feasible.

X-linked Hypophosphatemic Rickets

This involves the *phosphate-regulating endopeptidase homolog, X-linked (PHEX)* gene on human chromosome Xp, which encodes the PHEX enzyme. This probably regulates the balance of phosphate in the body, which plays a critical role in bone growth and strength. Mutations in the *PHEX* gene, which lead to an inactive enzyme, cause low levels of phosphate in the blood. The symptoms are soft and flexible bone abnormalities, resulting in, for example, bowed legs. It is the most common form of rickets that runs in families and affects about 1 in 20 000 newborns. It is also known as vitamin-D-resistant rickets since it does not respond to treatment by ingestion of vitamin D. There are several forms and a number of genes involved in rickets. Males are more frequently affected than females. Other human disorders associated with an X-linked dominant disease include:

- Fragile X syndrome, another X-linked dominant disorder, features DNA mutations (Section 4.4.1 in Chapter 4) and genomic imprinting (Section 7.1.5 in Chapter 7).
- Retinitis pigmentosa (RP; Section 6.3.5): Only 5–15% of cases are inherited in an X-linked pattern. At least 20% of heterozygous females develop retinal degeneration (an indication of X-linked dominance).

6.3.3   Cattle Disorders

The X-linked dominant condition is rare in cattle. It is more deleterious than the recessive form because one mutant gene in a male embryo results in death. Possession of the mutant gene in female Holsteins causes streaked hairlessness, especially in the flanks.

6.3.4   X-linked Recessive Pedigree

The inheritance patterns shown in Figures 6.13B and 6.14B resemble closely those of X-linked dominant, except that the heterozygous female is a carrier and rarely displays the phenotype, although she passes the mutation to the next generation. The characteristics of the two inheritance types are, therefore, also very similar, bearing in mind the differences in the hemizygous female in both cases. This is illustrated by an actual classic family tree from the marriage of a carrier female with hemophilia B, Queen Victoria, and an unaffected male, Prince Albert, affecting the royal families of Europe (Figure 6.16).

6.3.5   X-linked Recessive Human Disorders

There are a few hundred X-linked recessive disorders known, nearly all on the Xq chromosome.

Hemophilia B

This is associated with the *coagulation factor IX* (*F9*) gene on chromosome Xq, which encodes the protein coagulation

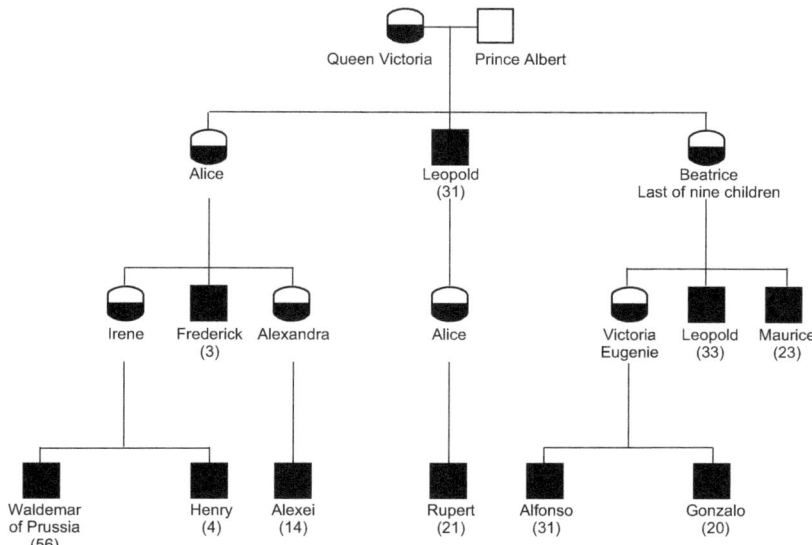

Figure 6.16   Part of the Queen Victoria family tree, showing only persons with one mutant hemophilia allele. Those in the family tree without the disease are not included. Half-black circles are female carriers not showing the disease. Full-black squares are affected males with their ages at death. The German, Russian and Spanish royal families, but not the British family, were affected.

factor IX. This is one of a group of related proteins essential for the formation of blood clots, thus preventing excessive blood loss after injury. More than 900 mutations have been associated with the *F9* gene, mostly involving single base-pair changes, as well as some deletions or insertions. Females are carriers when they have one altered copy of *F9* in each cell. The heterozygous female usually produces sufficient protein (about one-half of the normal amount) and rarely have two altered copies, so that females seldom have hemophilia B. In males, one copy of the altered protein causes the disease. It is a hereditary bleeding disorder, which affects about 1 in 20 000 newborn males worldwide. Recent studies used massively parallel sequencing to sequence *F8* and *F9* in the bones of Alexei (born in 1904, murdered in 1918; see Figure 6.16). He had only DNA that contained a nonsense mutant pattern associated with the *F9* gene, and therefore was a bleeder with hemophilia B and not hemophilia A, which most commonly involves a marked translocation mutation (Section 4.4.4 in Chapter 4). The DNA from his mother, Alexandra, contained both the mutant and the wild-type allele, and she was therefore a carrier, see Figure 6.16. Other rare but well characterized diseases include the following.

Ocular Albinism

The operative *G-protein-coupled receptor* (*GPR143*, also known as *OA1*) gene on the Xp chromosome, which encodes a protein that is important in the implementation of pigmentation in eyes and skin. All types of mutations, missense, nonsense, frameshift and splice-site, are represented. Most of the more than 60 mutations reported alter the shape and size of the protein. This prevents it from reaching melanosomes (Section 7.3.9 in Chapter 7) and controlling their growth. This leads to ocular albinism (OA), with light-colored eyes and permanently impaired sharp and stereoscopic vision. The color of the skin and hair is not affected (see other forms of albinism; Sections 7.3.3 and 7.3.7 in Chapter 7). An altered copy of the gene causes the disease in males but has little effect in females. OA1, the most common form, occurs in about 1 in 60 000 males and much less frequently in females, who may, however, be carriers.

Duchenne and Becker Muscular Dystrophies

The operative *dystrophin* (*DMD*) gene on Xp encodes the protein dystrophin, which is part of a group of proteins that help strengthen and protect muscle fibers. The *DMD* gene consists of 79 coding exons and stretches over 2.2×10^6 bp, representing about 0.1% of the entire human genome. It takes about 16 hours for polymerase II to transcribe this gene! The over 1000 mutations of the gene are more numerous than usual because of the increased likelihood of damage to such a large gene. These include point mutations, duplications and deletion of one or more exons, which 60% of sufferers have. Mutations causing Duchenne disease prevent any functional protein production. Other mutations in the same gene lead to the Becker form and may produce an abnormal, but still somewhat functional, protein. These are two related conditions, primarily causing weakness and wasting of the heart and muscles. Duchenne is the most common and severe form of muscular dystrophy. Together, the two forms affect about 1 in 3500 newborn males but very few females. Attempts using drugs to circumvent the various types of mutation are currently being investigated (see Section 8.5 in Chapter 8). The *dystrophin* gene is highly conserved. Homologs have been identified in mammals, birds and fish, as well as the worm *Caenorhabditis elegans*.

Retinitis Pigmentosa (RP)

RP is a group of eye diseases affecting the retina that causes progressive vision loss and, often, legal blindness. It affects 1 in 4000 people.

A number of genes and many mutations are associated with the disease. Autosomal dominance and recessive patterns, as well as X- and Y-linked patterns, are shown. Only 10% of cases of RP are X-linked. Around 90% of X-linked RP cases arise from mutations in the *retinitis pigmentosa GTPase regulator* (*RPGR*) gene, which encodes a protein required for normal vision. For males, one copy of the mutation is sufficient to cause the disease, which is generally severer than with the female, for which two copies of the mutation are generally necessary and is rarer. However, 20% of females with one mutated copy develop retinal degeneration and vision loss, and the ailment may be classified as partial dominant. Potential success in the treatment of RP has been demonstrated by gene therapy using CRISPR/Cas9 genome editing (Section 8.4.10 in Chapter 8). A patient with a simple $G > T$ mutation in the *RPGR* gene causes an aggressive form of X-linked RP. Stem cells from the skin are reprogrammed to become induced pluripotent stem cells. CRISPR editing is used to pinpoint and repair the defective mutation, which resulted in 13% of corrected stem cells. There is every likelihood that the corrected stem cells could be transformed into healthy retinal cells and be transplanted back into the patient to treat vision loss, but the final step is not yet allowed.

Some other disorders that conform to the X-linked recessive pattern include:

- **Lesch–Nyhan syndrome.** This is a hereditary disease caused by an enzyme deficiency, resulting in the overproduction of uric acid in the joints, which can cause gout. There are a variety of types in the reported more than 200 mutations, involving the relevant *hypoxanthine phosphoribosyltransferase 1 (HPRT1)* gene on the Xp chromosome. It is believed to have affected James I of England.

- **Fabry disease.** The operative *galactosidase, alpha (GLA)* gene encodes the enzyme alpha galactosidase, which is important in the recycling of old red blood cells. There are a number of types of mutation in the some 400 associated with the disease that produce an abnormal version of the enzyme. The mutation leads to skin lesions due to thin-walled blood vessels covered by a cap of non-malignant warts (angiokeratoma). It affects approximately 40 000 males but, again, many fewer females.
- **X-linked severe combined immune deficiency syndrome.** See Section 8.4.7 in Chapter 8.

6.3.6   Hemophilia in Animals

Hemophilia occurs in mice, cats, dogs, horses and other animals. Hemophilia B in mice has been corrected by gene therapy. Hemophilia B has recently been characterized in two unrelated domestic cats. The coagulation factor IX in the cat differs by only a few amino acids from the analogous protein encoded by the human and dog genes. The DNA sequence for the entire coding region of the *F9* gene has been determined for a healthy cat and compared with those for the affected cats. Two separate, single nucleotide changes in the *F9* gene have been observed in the affected cats. In one cat, there is a change from GGA (Arg) to TGA (stop), leading to a truncated protein. It is one of the most common types of mutation in human hemophilia B. In the other cat, a mutation, TGT (Cys) to TAT (Tyr), results in a detrimental structural change. Hemophilia is one of the few sex-linked traits in dogs. It arises, as in other animals, by spontaneous mutations, six of which have been identified. Once hemophilia appears, the defect can be transmitted through many generations. The mutations run the whole gamut, *i.e.*, single base change, deletions and insertions. In affected Labrador Retrievers, an entire gene is deleted! The severity of the condition is variable and some affected male dogs show only limited symptoms.

Horses with hemophilia rarely live long. As with humans, the condition is usually found only among males.

6.3.7 Y-linked Pedigree

There are two situations where the inheritance pattern is very straightforward, namely in Y-linked and mitochondrial DNA traits. Both Y chromosomes and mtDNA are copied and passed down virtually unchanged through generations (barring naturally occurring mutations). A typical family tree depicting the Y-linked pedigree is shown in Figure 6.17.

Characteristics of the pedigree include:

- Only males have a Y chromosome, so the Y-linked traits can only be passed from male to male. No female is ever affected (open circles in Figure 6.17).
- The Y chromosome undergoes little or no change when passed from parent to child.
- The Y chromosome trait is hemizygous. That means it is expressed as if it were dominant since there is no offsetting allele.

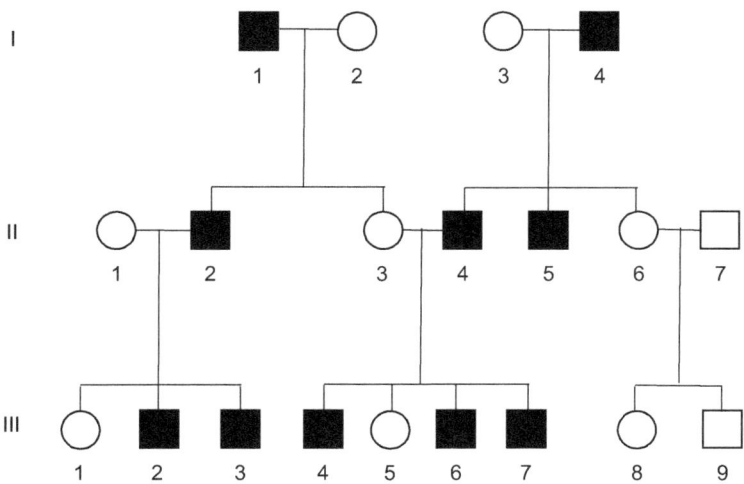

Figure 6.17 A typical representation of Y-linked inheritance of traits. Open circle, unaffected female. Open square, unaffected male. Black square, affected male.

- All Y-linked traits show 100% penetrance but variable expressivity.
- The Y chromosome is small and contains relatively few genes. Genes are mainly found in the p-arm of the X chromosome of the male and are usually related to male functions. Mutations in these genes usually result in infertility, and therefore there are no offspring. Only about 40 Y-linked traits have been discovered and even these are extremely rare.

6.3.8   Human Disorders: Testicular Disorder of Sex Development (46,XX)

The operative *sex-determining region (SRY)* gene is only present on the Yp chromosome. It encodes a protein important in transforming the mammalian embryo into a fetus with male gonads. Unless the SRY gene becomes activated, the embryo is programmed to become a female. In most cases of testicular disorder, the gene is transferred from the Y chromosome to an X chromosome (translocation; Section 4.4.3 in Chapter 4). This is usually during the formation of sperm cells in the father of the affected person. Since the fetus still has the *SRY* gene, albeit on an X chromosome, it will still develop as a male, even without a Y chromosome, but will have small testes and sterility, as well as other abnormal characteristics.

- Syndromes 47,XXY and 47,XYY, which are only associated with males, are mentioned in Section 5.2.4 in Chapter 5.
- A rather novel human disorder, which may have a Y-linked pedigree, is hairy ear rims, which are common in parts of India. In one family group, all 13 men, but no women, were affected. It might, however, be a sex-linked trait (Section 7.1.2 in Chapter 7).

6.3.9   Richard III

The Y chromosome DNA is copied and passed down virtually unchanged from the father to his son(s) and their son(s) and so on. It was hoped to use this relationship to further substantiate the identity of the skeletal remains that were almost certainly Richard III (see Section 6.4.1). Richard III had no children, but a male line could be established through his great-great-grand-father, John of Gaunt, who has descendants to the present day through a succession of male children. It was found that the Y DNA of the living males did not match that of Richard III. This was attributed to false paternity somewhere along the long line rather than casting doubt on the identity of the skeletal remains. DNA phenotyping (Section 8.1 in Chapter 8) on the remains of Richard III indicated a 96% probability of him having blue eyes and a 77% probability of boyhood blond hair. These characteristics are consistent with a portrait believed to be an accurate one of the monarch.

6.4   MITOCHONDRIAL INHERITANCE

Some properties of mitochondria are described in Table 1.1. An egg cell rejects sperm mitochondria, and therefore the egg cell almost only contains mitochondrial DNA provided by the female (Section 5.3.2 in Chapter 5). The inheritance pattern is therefore simple (Figure 6.18), as we found with paternal inheritance provided by the male Y chromosome.

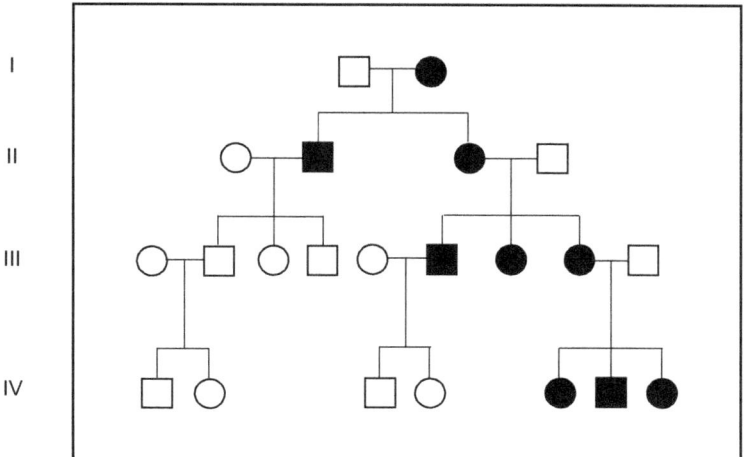

Figure 6.18   Possible inheritance of a mitochondrial disorder through four generations. Affected persons are shown with blocked symbols. Only mothers pass on the condition to their children.

Mothers can pass their mitochondrial DNA to their children, but only their daughters can pass it on to their children and, barring the occasional new mutation, their mtDNA sequences should be identical. Males and females are equally affected if enough people are considered.

In Figure 6.18, the great-grandmother (generation I) has a mitochondrial disorder. Both of her children in generation II have the mitochondrial disorder. Only the girl passes this on to her three children in generation III. Once again, only one daughter in generation III passes this on to her three children in generation IV.

6.4.1   Richard III (Again)

A striking example of the value of mitochondrial inheritance in tracking descendants is afforded by the story of Richard III. An examination of the mitochondrial DNA in the teeth and bones of the skeletal remains, almost certainly those of Richard III, has gone a long way to confirm the King's identity. Mitochondrial DNA is only passed through the female line and rarely mutates, so it was necessary to find a female relative to Richard III and also to find a direct descendent from her to the present day. Fortunately, and remarkably, this was possible (Figure 6.19). Richard III and his sister, Anne of York, would have identical mtDNA inherited from their mother, Cecily Neville. Anne has living descendants. The mtDNA of (15 great)-grandson Michael Ibsen is perfectly matched with the sequence of that of Richard III. The mtDNA of Wendy Duldig (17 great)-granddaughter differs by only a single base from that of Richard III. The possibility of identical or near-identical mtDNA arising by chance is almost certainly ruled out since the type of mitochondrial sequence is a very rare one.

6.4.2   Human Disorders

A cell with a uniform collection of mitochondria (homoplasmy; Section 1.7.4 in Chapter 1) may contain completely normal or completely mutant DNA. However, two distinct populations of the normal and mutant DNA may coexist in the cell (heteroplasmy). If the mutation is associated with a disease, the larger

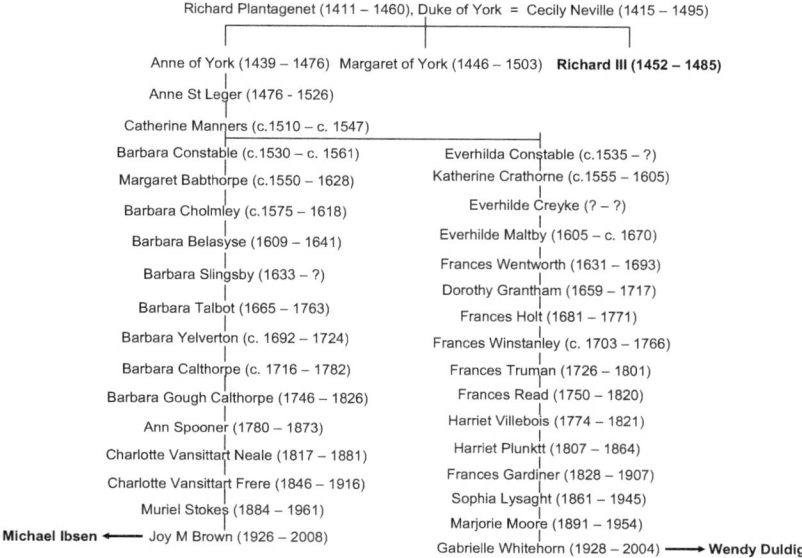

Figure 6.19   The family tree of Anne of York showing the maternal inheritance to the present day.

the proportion of mutant DNA the more likely it is that this person will show the disease, and penetrance and expressivity will vary. Mitochondrial diseases arise from the same range of mutations as are seen in mutations involving nuclear DNA. Recalling that mitochondrial DNA only contains 37 genes, it is easy to understand that the associated disorders are relatively few in number, although many are deadly and mainly affect children. It has been estimated that 1 in 4000 people are affected by mitochondria diseases. They are often involved with muscular and neurological disorders because the mitochondria produce more than 90% of the energy supply for the body. Eye disorders, in particular, are involved, such as:

- **Leber hereditary optic neuropathy:** Four variants of the mitochondrial-encoded *NADH:ubiquinone oxidoreductase* (*MTND*) gene in various parts of mitochondrial DNA (Figure 1.14) give rise to four forms of the protein NADH dehydrogenase. This enzyme is part of a complex involved in oxidative phosphorylation. This is vital in the conversion of ADP into ATP, which, on hydrolysis, releases the energy necessary to support many biological processes, such as muscle contraction, the synthesis of proteins, nutrition, *etc.* Mutations involving different single amino acid changes in any of the proteins lead to Leber hereditary optic neuropathy. More prevalent with males, it usually begins in a person's teens, and visual loss and rapid blindness within months is incurred.

- **Kearns–Sayre syndrome:** Most people with Kearns–Sayre syndrome have a single, large deletion, often involving 4997(!) nucleotides. This deletion results in a loss of 12 mitochondrial genes required for implementing oxidative phosphorylation. There is a progressive eye problem (movement of the eye) and even loss of vision. It is often accompanied by cardiac and balance problems. The attempted replacement of faulty mtDNA in females by normal mtDNA through gene therapy has been sanctioned in the UK (Section 8.4.2 in Chapter 8).

Deviations from an Expected Phenotype

We have had many examples of a straightforward link between the genotype and the phenotype. However, deviations from this relationship abound.

7.1 WHEN THERE IS AN OVERRIDING ROLE OF ONE PARTNER

These involve genes present in males and females. They are expressed differently depending on the sex of the owner. The traits may be dominant in one sex but recessive in the other. This is termed sex-influenced genes (traits). In contrast, sex-limited traits, also autosomal, are expressed only in one sex, *i.e.*, the trait has zero penetrance in one sex. Sex-linked traits involve the sex chromosomes and were dealt with in Section 6.3 in Chapter 6.

7.1.1 Sex-influenced Traits

Sheep Horns

Dorset sheep, both male and female, are horned (allele a), whereas Suffolk sheep are hornless (polled, allele b). When a Dorset (horned) sheep (aa) of either sex is crossed with the appropriate Suffolk (polled) animal (bb), the progeny (ab) are found to be horned males and polled females. From previous considerations, we might expect a 3 : 1 ratio of horned : polled or a 3 : 1 ratio of polled : horned, depending on which is the dominant gene, irrespective of their sex. When these progeny (ab) are mated, the resulting females segregated as one horned to three polled, while the males appeared in a three-to-one horned-to-hornless ratio (Figure 7.1).

The results show that the genotypes aa and bb are horned and polled, respectively, in both sexes and only the expression of the heterozygote animal (ab) is influenced by its sex. The horned character is thus dominant in the male and recessive in the female. This behavior is ascribed, in part, to differences in the

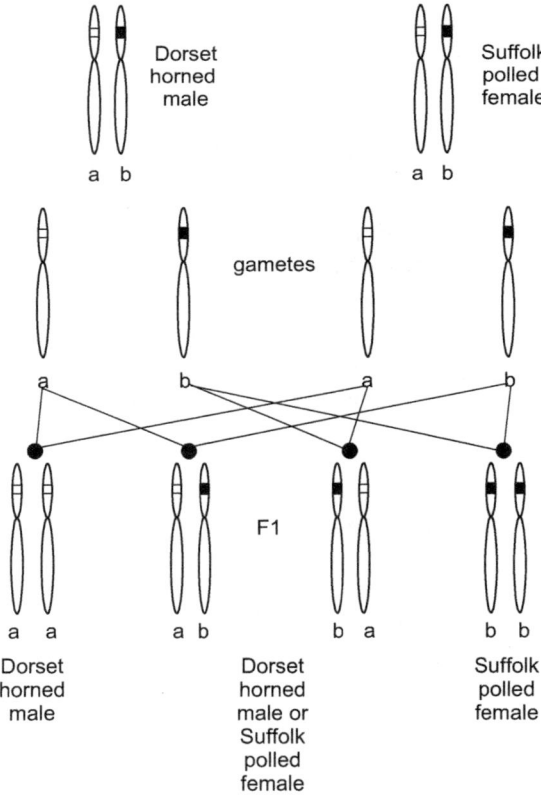

Figure 7.1 The resulting phenotypes from crossing Dorset with Suffolk sheep.

sex hormonal environments in the male and female. These sex-influenced traits are, therefore, largely found in higher animals with well-defined endocrine systems (ductless glands involved with hormones).

Human Baldness and Other Human Conditions

Similar considerations are also believed to apply to early (20–30 years) male-pattern baldness (other types of baldness are known). The baldness gene is dominant in males but recessive in females. Therefore, a heterozygote man, but not a woman, with the one gene for baldness will probably lose his hair, while a woman needs two balding genes to be bald and this is rare. Other human conditions believed to be sex influenced and more common in males include harelip and cleft palate, gout and stuttering.

7.1.2   Sex-limited Traits

These conditions are a little more extreme than the previous ones, as now particular autosomal genes can *only* be expressed in a particular sex. Although the genes responsible for elk antlers and peacock plumage are carried by both genders, they are expressed only in the male. Both men and women have genes for beard and mammary gland development, and, yet again, they are associated virtually completely with the male and female, respectively. This dominance of one sex in the phenotype is due, in part, to the different cellular environments in males and females, contributed by sex hormones.

7.1.3   X-inactivation

This phenomenon is unique to mammals. Females have two X chromosomes, one originating from the mother (X_m) and one from the father (X_p). The female embryo needs to muzzle one of its two X chromosomes to prevent an overdose of X chromosomes, compared with the only one X chromosome of the male. This is termed dosage compensation. It is a regulatory mechanism that enables a gene phenotype on an X chromosome to be equally expressed in the XY male and XX female. About 15% of all X-linked genes avoid inactivation and these are transcribed at higher levels in women than in men. Early in mammal development (at about the 20-cell stage), one of the two X chromosomes, X_m or X_p, in female somatic cells will become inactivated, but which of the two is chosen is completely random. Once chosen, all descendent cells will have the same X chromosome inactivated. Inactivation is accomplished by extensive methylation of the histones associated with the inactivated X chromosome. This is termed an epigenetic modification (Section 7.4.2). The randomness of the inactivation means that the female is a mosaic in which some cells express the maternal X chromosome and some cells express the paternal X chromosome, Figure 7.2.

The inactivated X chromosome in females becomes a blob of chromatin attached to the nuclear membrane and, when stained, is visible under a microscope at the interphase stage of the cell cycle. No such inactivation occurs in the solitary X chromosome in males. Regardless of the occurrence of mosaicism, females will usually show the phenotype of the dominant, normal allele because enough cells will be present in the body to express the dominant, normal trait. Furthermore, the motley mosaic of cells is usually "invisible" in the female. A couple of exceptions to this generalization come from the human and feline worlds.

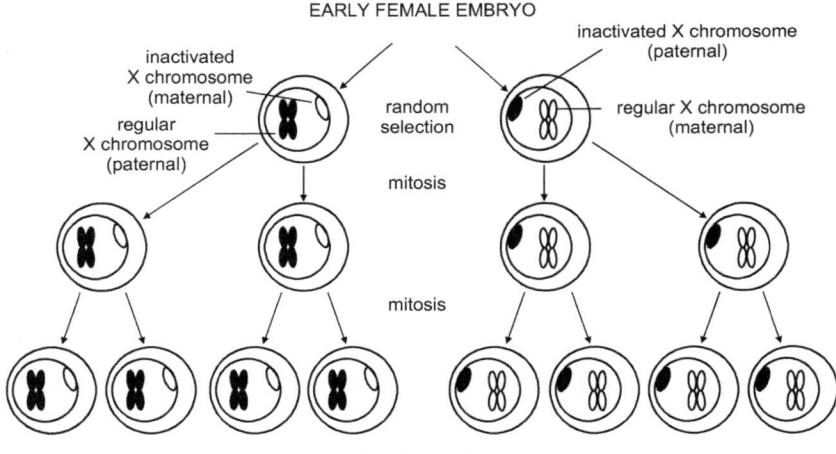

Figure 7.2   The generation of a mosaic female cell. Black and white blobs represent inactivated X chromosomes (Barr bodies). The complete organism is a mosaic of cells expressing alleles from either the active X_m or X_p chromosome.

Anhidrotic Ectodermal Dysplasia

The *ectodysplasin A* (*EDA*) gene on chromosome Xq encodes an ectodysplasin A protein, which is important in developing hair, teeth and the sweat glands. There are more than 80 different types of mutation in the *EDA* gene, causing a nonfunctional version of the protein. This is the basis of the X-linked form of the disease, which accounts for 95% of all cases of the disease. Most people with this condition have a reduced ability to sweat because of fewer or nonfunctional sweat glands. This is serious, since perspiration is crucial for body cooling. Worldwide, 1 in 17 000 people are affected. It is more common in males, since only one copy of the gene in each cell causes the condition, leading to a completely sweat-defective body. In females, the mutation must be present in both copies of the gene to cause the condition. Females with one altered gene are carriers and have mild symptoms. This heterozygote female shows a mosaic pattern, in which parts of the body sweat normally (normal allele; mutant allele inactivated) and other parts do not sweat at all (mutant allele; normal allele inactivated). The condition is X-linked recessive (Section 6.3.4 in Chapter and Figure 6.14B) for a carrier mother and an unaffected father.

A novel test is used to assess mosaic anhidrotic dysplasia in females. The back of the subject is painted with a dried iodine/starch mixture and exposed to a very warm environment. The active sweat pores produce sweat droplets, which provide the necessary medium for the iodine to react with the starch to give the characteristic blue color. The mosaic condition will show up as an uneven distribution of minute blue (dark) spots.

Tortoiseshell Cats

One of the several genes controlling fur color in cats, as well as mice and other animals, is located on the X chromosome. There are two alleles, one directing an orange (B) color and one directing a black (b) color. When we consider male cats, there is no ambiguity. They have only one X allele for a particular trait (in this case, coat color), a condition termed hemizygosity. Their one inherited allele either means the cat will be orange if it has the B allele or black if it has the b allele (Figure 7.3). The male Y

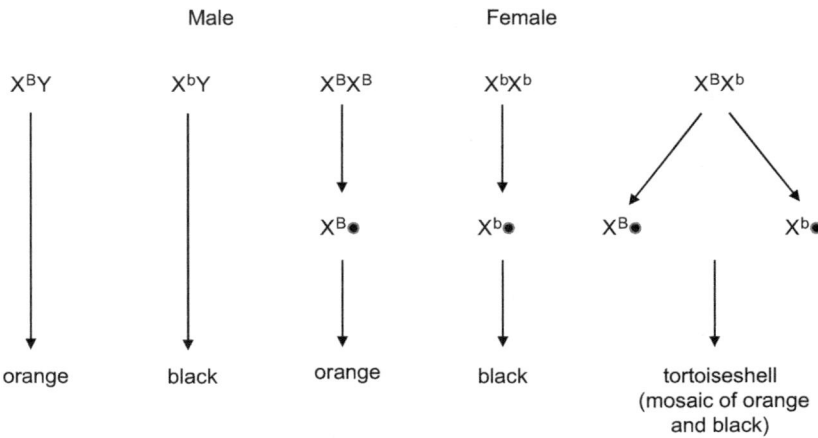

Figure 7.3 The expression of orange, black and tortoiseshell coat colors in male and female cats. The inactivated X chromosome is represented by a dark circle.

chromosome makes no contribution, not having this particular color gene.

With homozygote females, the cat will either be orange (BB) or black (bb), even though it loses one of the X chromosomes (alleles) by inactivation since it gains the corresponding X chromosome from the male. With heterozygote female cats (Bb), some complexity arises. Arbitrary X-inactivation of either B or b alleles means the female skin tissue cell will end up as a patchwork of black or orange colors, respectively, *i.e.*, a tortoiseshell mosaic results (Figure 7.3). A calico cat is a tortoiseshell cat with an additional gene, giving non-pigmented (white) patches, termed piebald.

7.1.4 Genomic Imprinting

Genomic imprinting is unique to placental mammals (and flowers). Normally, there are two working copies of a gene in an animal chromosome, one contributed by the mother and the other by the father. About 1% of the genes in the total DNA (genome) of a mammal are, however, imprinted—in humans, this is approximately 100 genes, with many more suspected. This means that only one of the genes is working. That one functional gene may originate from either the mother or the father, as with X-inactivation (Section 7.1.3). Imprinted genes are usually confined to only a few of an animal's autosomes; in humans, there is a noticeable cluster of imprinted genes in chromosomes 11p and 15q. The process begins during gamete formation. Certain genes are imprinted in males in their developing sperm. In females, other genes are imprinted in their developing egg. These imprinted genes from the father and the mother will persist in the offspring, except for those cells that make gametes, where the imprints in those cells are erased. Genomic imprinting is promoted by methylation of DNA during egg or sperm formation. The epigenetic marks (methyl groups) persist during the life of the organism but are reset during egg or sperm formation (Section 7.4).

Dwarf Mice

The concept is well illustrated in the early studies involving the *insulin-like growth factor 2* (*IGF2*) gene on chromosome 7 in mice (Figure 7.4).

This gene encodes the protein IGF2, which is vital in the growth and development of a fetus, in which the gene is most highly active. Two copies of a mutated gene (homozygous mutants) lead to a dwarf mouse. Two copies of the normal gene (ignore the imprinting) lead to a normal-size mouse. The big difference from previous considerations arises when we consider the heterozygote. Without imprinting, there would be a ratio of three-to-one of normal-to-dwarf mice, where the normal gene is dominant. Now, with imprinting, the normal and mutant genes from the father are expressed, but the normal allele from the mother is imprinted, *i.e.*, silenced and non-expressed (Figure 7.4). This means that a heterozygote that receives the one normal allele from a father is a normal size, whereas a heterozygote that receives the one normal imprinted allele from a mother is a dwarf. It is as if the imprinted allele did not exist. So, the normal *IGF2* gene is imprinted to function poorly during egg production but to function normally through sperm-producing tissues. As might be expected, things are not as simple as this. The *IGF2R* gene on chromosome 17 in mice encodes IGF2R, which is a receptor that mops up the IGF2 protein and prevents it from promoting fetal growth. The *IGFR2* gene is also imprinted, but now it is expressed from the maternal and not the paternal copy of the gene. In this way, a balance in the size of the fetus is

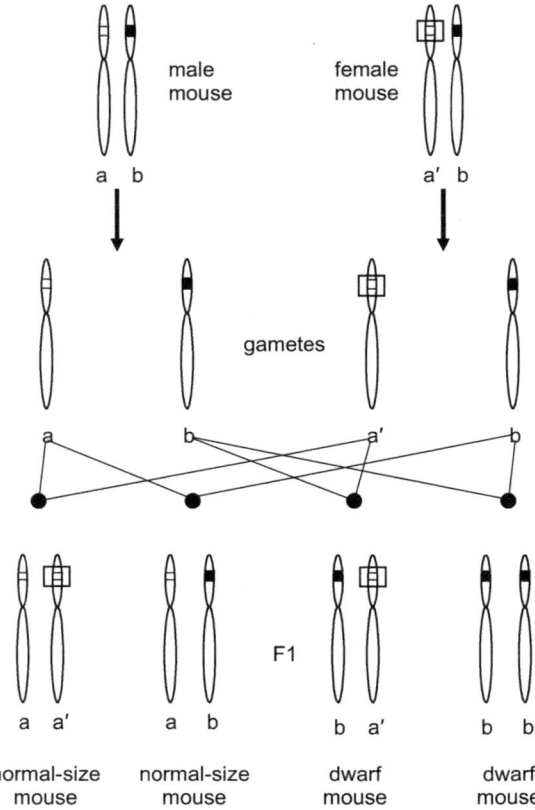

Figure 7.4   The results of imprinting on the size of mice. The imprinted gene a′ is enclosed within the square.

achieved. The *IGF2* gene is one of the most important fetal genes and has been investigated in a number of animals, including mice, cattle and pigs, as well as in humans.

7.1.5   Uniparental Disomy

Almost invariably, one of the autosomal pair of chromosomes in the nucleus originates from the mother and the other from the father (heterodisomy). Only very occasionally are both chromosomes inherited from the same parent. This is termed uniparental disomy (UPD) and is a special instance of genomic imprinting. There are a variety of ways in which UPD may arise, the most common of which involves nondisjunction in mitosis or meiosis. Fertilization of a disomic egg by a normal sperm leads to a trisomic fetus (see Figures 3.14 and 5.5). If a parental chromosome is then lost (trisomic zygote rescue), then UPD will result in one-third of cases, Figure 7.5A.

This is the most common mechanism for UPD. Alternatively, fertilization of a nullisomic egg by a normal sperm leads to a monosomic fetus. If this is duplicated as a post-zygote event (monosomic zygote rescue), UPD results, as shown in Figure 7.5B.

7.1.6 Human Disorders

UPD rarely leads to health problems because there are still two chromosomes in the nucleus with the desired genes, even if they arise from only one of the parents. However, in the very unlikely event that the one parent shows UPD involving genetically im-printed essential genes, then this can lead to clinical disorders. Few have been reported.

Prader–Willi Syndrome (PWS) and Angelman Syndrome (AS)

These are genetically related autosomal dominant diseases. PWS and AS are clinically distinct but have similar characteristics. Obesity and neurological disorders are more severe in PWS, while mental retardation and impaired speech are severe in AS. Both arise from loss of DNA segments involving one or several genes in nearby regions of human chromosome 15q11-13. These are essential for normal development. They were the first iden-tified imprinting disorders.

- **Prader–Willi syndrome:** Genes in this particular region are normally expressed only from the father. The genes from the mother are inactive (imprinted). If, therefore, the material donated from the father is compromised in some way, or if both chromosomes originate from the mother (UPD), then PWS is experienced. The majority of cases of the syndrome arise from deletion ($3-4 \times 10^6$ bp) or another mutation of the relevant genes from the paternal chromosome (70%). If the person has two maternally derived chromosomes 15 (maternal UPD), about 25% of the cases of PWS result and

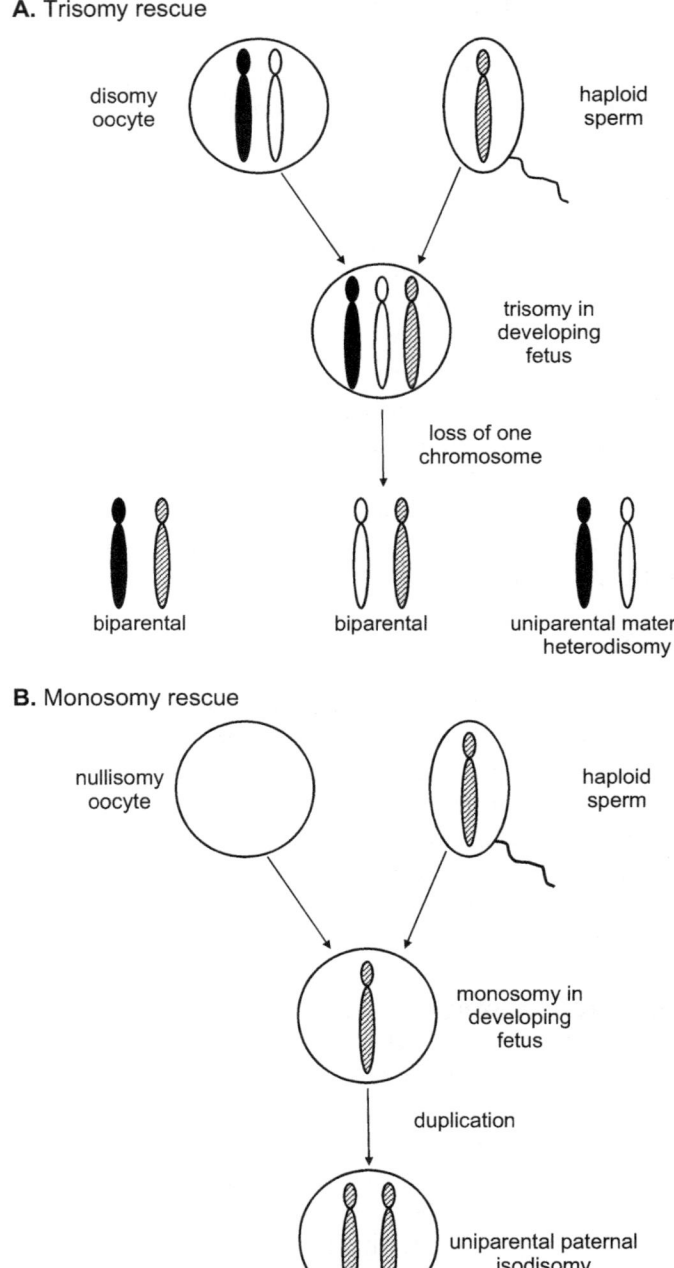

A. Trisomy rescue

disomy oocyte

haploid sperm

trisomy in developing fetus

loss of one chromosome

biparental   biparental   uniparental maternal heterodisomy

B. Monosomy rescue

nullisomy oocyte

haploid sperm

monosomy in developing fetus

duplication

uniparental paternal isodisomy

Figure 7.5 (A) Trisomic and (B) monosomic zygote rescues. In these examples, maternal and paternal UPD in human chromosome 15 are shown, leading to Prader–Willi syndrome (PWS) and Angelman syndrome (AS), respectively.

are usually associated with advanced maternal age. About 5% of cases arise from imprinting errors, Figure 7.6.

- **Angelman syndrome:** Genes in the same region are this time only expressed from the mother. Those from the father are imprinted. In addition, unlike PWS, only certain areas of the brain, and not other tissues, are affected. If the maternal copy is lost by deletion or another mutation, AS results, as is the case for 80% of sufferers. A few percentages of cases arise from paternal UPD, resulting from inheritance of two imprinted (silenced) copies of chromosome 15 from the father. The state of the chromosome pairs in these diseases is summarized in Figure 7.6. Most cases of PWS and AS arise *de novo*, as random events during the formation of egg or sperm cells, or in the early development of the embryo. Epigenetic factors play an important role in the incidence of these diseases (Section 7.4.3).

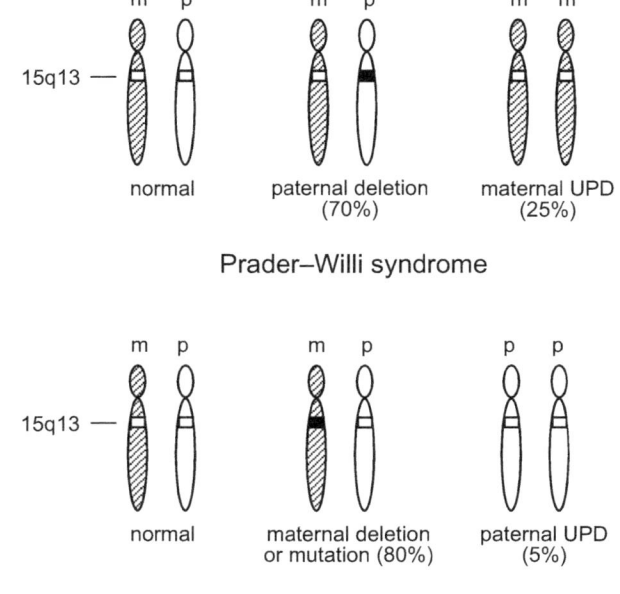

Figure 7.6 Chromosome pairs in Prader–Willi and Angelman syndromes, and their approximate occurrences in the diseases. A very small percentage arise from unknown causes.

Genomic imprinting has been linked to a few other diseases. The higher frequency and severity of fragile X syndrome (Section 4.4.1 in Chapter 4) in males may be ascribed to imprinting by the mother. When inheriting Huntington disease (Section 4.4.1 in Chapter 4) from the father, an earlier onset of the disease is observed, suggesting a genomic imprinting role.

7.2   THE TEMPERING OF THE IMPACT OF MUTATIONS

We now have to consider some modifying effects (good and bad) of traits or disorders. How much of the mutant phenotype is expressed (expressivity)? In other words, is there a range of appearances or symptoms originating from the genotype? Then, what is the likelihood that a person will experience the effects of the disorder when they have the mutant gene (penetrance)? Finally, how will the severity and age of onset of a hereditary disease change, if at all, in succeeding generations (anticipation)?

7.2.1   Expressivity

Although two animals may have an identical, unusual genotype and the same associated genetic condition, their phenotypes, which might have been expected to be identical, may vary in the severity of a disorder or the degree of physical change. The extent to which the phenotypes of the animals differ is termed the expressivity. Variable expressivity may be largely explained by the presence of epistasis genes (Section 7.3.6). The phenomenon is displayed in humans and animals.

7.2.2   Human Disorders: Expressivity

Marfan Syndrome

The *fibrillin-1* (*FBN1*) gene on the q-arm of human chromosome 15 is responsible for synthesizing a large protein called fibrillin-1, which is important in the production of connective tissue for supporting joints and organs in the body. Over 1300 different mutations (mostly missense) have been reported. One altered allele of the two in the gene causes Marfan syndrome. Mutations cause either a nonfunctional or a reduced amount of protein, leading to a disorder of the connective tissues that affects 1 in 5000 humans worldwide and was first described more than 100 years ago. Most people with the syndrome show some external evidence of the disorder (nearly 100% penetrance), but there is variable expressivity. This means there are a wide range of symptoms, which depend apparently on the position of the mutation in the chromosome. Sometimes, a person will carry the mutant gene but not develop the disease. A person (male or female) may have mild symptoms, being tall and thin and have long fingers. Another person with the syndrome, on the other hand, may have life-threatening heart and blood vessel failures. Most people with the faulty gene inherit it from an affected parent, but about one-quarter do not have a parent with the condition. This is probably because the *FBN1* gene mutates for the first time in the egg or sperm of the parent and is passed on to the child by this means.

Neurofibromatosis Type 1

The one gene involved, *neurofibromin 1* (*NF1*) is on the long arm of chromosome 17 and encodes the neurofibromin protein (containing almost 3000 amino acids), which is produced in many cells. This regulates cell growth and suppresses tumor formation. Mutations lead to the type 1 form of neurofibromatosis, which accounts for 90% of sufferers. It is one of the most common single-gene disorders affecting the human nervous system, showing a usually benign growth along nerves in many parts of the body. It occurs in about 1 in 3000 humans worldwide. Over 1000 different mutations have been associated with the disorder, many of which encode a less effective, shortened version of protein. One copy of the altered allele is sufficient to promote the disease, but two altered copies must be present in each cell to trigger tumors. The disorder is highly penetrant, *i.e.*, nearly everyone with that genotype will contract the disease (Section 7.2.4), but there is a large expressive variability (Section 7.2.1). Some sufferers may have just a few spots, while others may have gross disfigurements, as with the famous "elephant man". The reasons for this variability are uncertain. As with Marfan syndrome, persons may inherit the disease from an affected parent (about one-half) or may acquire the mutation *via* the gametes of an unaffected parent or early in the development of the embryo.

7.2.3   Animal Disorders

Polydactyly

Polydactyly is the occurrence of more than the usual number of toes in an animal and is especially associated with cats. Most domestic cats have five toes on each front paw and four on each hind paw. With polydactyl cats, as many as eight toes (usually on the front paw) have been noted. A very early but comprehensive study of all possible mating routes involving polydactyly (poly) cats and non-polydactyly cats established that the polydactyly allele (A) in the *Pd* gene was dominant to the non-polydactyl allele (a). Mating two poly cats (Aa) gave a 3 : 1 ratio of poly : normal kittens. Further, there was no reduction in the size of the litter compared with those obtained with two non-poly cats. This means that the homozygous polydactyly state was not lethal (compared to the yellow mice discussed in Section 6.2.4 in Chapter 6). There is a fair degree of expressivity and incomplete penetrance. Thus, the number of extra toes in poly cats varies widely and not every cat with the mutant *Pd* gene showed the unusual characteristic. The phenomenon has been observed in a number of animals, including humans.

In humans, polydactyly is observed in the hands and feet, often only as a mild effect. It arises from mutations in a gene on human chromosome 7p and affects 1 in 500 people.

7.2.4   Penetrance

If everyone who carries a mutant gene shows the corresponding mutant characteristic or disorder, that gene is said to show complete (100%) penetrance. Usually, however, a number of people (even perhaps a few out of very many) can escape the disorder even though they carry the mutation. The disorder is then said to show reduced or incomplete penetrance, expressed as a percentage. Most inherited autosomal dominant mutations, like Marfan syndrome and neurofibromatosis, have very high, if not complete, penetrance, although they have widely varied expressivity.

7.2.5   Human Disorders: Penetrance

The degree of penetrance may depend on the mutation causing the disorder. In cystic fibrosis (Section 4.3.7 in Chapter 4), for example, which is an autosomal recessive inherited disease, the most frequent mutation in the *CFTR* gene on chromosome 7q is Delta F508 (a deletion of phenylalanine at position 508 on the DNA) and the associated penetrance is 99%. An R117H mutation, on the other hand, has a small percentage penetrance, and therefore very few with the mutation suffer. The features responsible at a molecular level for the low penetrance are unknown.

BRCA1- *and* BRCA2-*Induced Breast Cancer*

Breast cancer 1 and *2, early onset* genes (*BRCA1* on human chromosome 17q and *BRCA2* on human chromosome 13q; *BRCA* denotes *b*reast *ca*ncer) encode proteins BRAC1 and BRAC2, respectively. These, together with other proteins, repair double strand breaks in damaged DNA and function as tumor suppressors. They help to keep breast cells growing normally. Over 1000 mutations have been identified in the two genes. Most mutations in the two genes lead to an abnormally short or absent protein. These are unable to help repair damaged DNA, leading to uncontrollable cell growth and division and incipient tumor formation. *BRCA1* and *2* gene mutations are associated with an increased risk of hereditary breast cancer, as well as other types of cancer. They account for 5–10% of all breast cancers. There is a high penetrance (see Section 7.2.4) for the genes: 65–80% up to 80 years of age. Men with abnormal *BRCA1* and 2 genes are more likely to develop prostate and other cancers than men with the normal gene.

The likelihood of a person acquiring two deleterious somatic mutations, both after conception (non-hereditary), are relatively low. However, a person receiving a germline mutation in one copy of either *BRAC* gene from one parent in all somatic cells just requires an additional mutation after conception in a single breast cell during the mitotic cell cycle for the initiation of a tumor (hereditary). This Knudson two-hit hypothesis appears applicable to many forms of cancer.

Osteogenesis Imperfecta (OI)

The main gene involved, *collagen type 1, alpha 1 (COL1A1)*, on human chromosome 17q produces part of a large protein, collagen type 1, which is important in building bone and strengthening muscle. There are many different mutations that cause the four types of OI of varying severity, from a mild form to the worst, being death at birth. This is a congenital disease (*i.e.* present or recognized at birth), causing extremely fragile bones and short stature, and was first documented in a 1000 BC mummy. About 1 in 20 000 persons worldwide have the disease. Most cases of OI are autosomal dominant with incomplete penetrance, a greatly varied expressivity and inherited from a parent. Some result from a new genetic mutation.

It is believed that reduced penetrance and variable expressivity result from a combination of reasons, including genetic and environmental, although they are far from understood. As we have seen, there are a number of steps between gene transcription and protein expression. Interference with any of these steps (epigenetic factors) could account for penetrance and variable expressivity. Penetrance and expressivity are often discussed jointly, since both are concerned with the way the phenotype may reflect the genotype. In fact, one can consider zero penetrance (no one has the mutant phenotype) as an extreme form of expressivity in which the range of expressivity includes non-expression.

7.2.6 Anticipation

Anticipation in a genetic condition is where signs and symptoms of a disorder become more severe and appear at an earlier age as the disorder is passed from one generation to the next. It is well chronicled in the case history of a family having myotonic dystrophy (Section 4.4.1 in Chapter 4).

Case History for Three Generations with Myotonic Dystrophy

- The grandfather, a 57-year-old man, had the onset of muscle weakness in his late 30s. He had 350 CTG repeats.
- The daughter had more severe muscle weakness, with onset as a teenager and had 520 CTG repeats.
- The grandson had prenatal onset of severe muscle weakness and, two weeks after birth, he died. He was found to have more than 3000 repeats.

Anticipation has been demonstrated or suggested in a number of human diseases. It is observed in many, but not all, trinucleotide repeat disorders (Section 4.4.1 in Chapter 4), often involving the nervous system.

7.3   ONE GENE/ONE PROTEIN VARIATIONS

The Maxim, "One gene encodes one protein and one phenotype" is a very approximate one and rarely encountered. Until now in the book, a straightforward relationship between gene and phenotype has generally been assumed. As might be anticipated, this is a relatively rare situation and other behaviors are more common. Thus, a single gene may be responsible for more than one protein (see Section 3.1.4 in Chapter 3), and therefore multiple phenotypes (pleiotropy). More than one gene may contribute to a single phenotype (polygenes). One gene may interfere with the operation of another gene that is attempting to carry out a particular task (epistasis).

7.3.1   Pleiotropy

In pleiotropy, a single gene can now be responsible for multiple phenotypes, which may be a human disease with a wide range of symptoms or an animal with a number of different forms of a physical characteristic. There are a number of reasons for pleiotropy. A single protein may still be produced but is able to promote multiple traits. A single locus may produce different mRNAs, which may arise from alternative splicing during transcription or because of alternative start/stop codons within the locus. These alternate transcripts will lead to different forms of the protein with altered functions and possibly leading to altered traits. The transcribed RNA can be further modified through mRNA editing. The cell can make nucleotide substitutions in the mRNA, leading to differences in the encoded protein.

7.3.2   Human Disorders

Pleiotropy is very common in a number of rare, but important, single-gene human disorders and its occurrence may give insight into the biological function of a specific gene.

- Marfan syndrome (Section 7.2.2): One gene is responsible for a number of phenotypes (symptoms), including thinness, limb elongation and increased susceptibility to heart disease.
- Phenylketonuria (PKU) is an autosomal recessive disease (Section 6.2.11 in Chapter 6). A defect in a single gene that encodes phenylalanine hydroxylase results in multiple symptoms associated with PKU, namely eczema, pigment defects and mental retardation.
- Hemoglobinopathy encompass all genetic diseases involving an abnormal production of hemoglobin and are the most common inherited diseases in the world. They include sickle cell disease and the milder sickle cell trait (Section 6.2.11 in Chapter 6). A person who has the trait will not contract sickle cell anemia and is less likely to get malaria. This is an example of an antagonistic pleiotropy gene, having both positive and negative effects.
- Waardenburg syndrome: This is a group of conditions passed down through families and inherited in an autosomal dominant pattern. One common form results from mutation of the *paired box* (*PAX3*) gene on human chromosome 2q, which leads to a number of pleiotropic effects, including colorless hair and skin, hearing loss, and unusual hand and facial features.

7.3.3 Animal Disorders

- Waardenburg syndrome also exists in a number of animals, including cats, dogs (especially Dalmatians) and ferrets. At least 40% of white cats with blue eyes are deaf, and a white cat with one blue eye and one yellow eye, for example, is only deaf on the blue-eyed side.
- Albinism: A single mutation in one of the enzymes catalyzing a stage in the tyrosine-to-melanin conversion (Figure 6.11) can cause multiple phenotypes. As well as albino skin, hair and eye pigmentation, defects in vision may also be present.
- Frizzle feathered chickens: Some chickens have feathers that curl outward rather than lying flat. Other phenotypic effects of the dominant "frizzle" gene responsible include

abnormal body temperature, higher blood flow rates and greater digestive capacity but a lesser egg-laying capacity than the wild-type counterpart.

7.3.4 Polygenes

When a phenotype is under the control of only one locus, only two distinct traits within the phenotype are observed (the wild type and the mutant form) with a dominant/recessive or codominant condition, or three distinct traits with a partial dominant situation (Section 6.2 in Chapter 6). This is termed discontinuous variation. If, however (and this is more likely), a phenotype is controlled by more than one gene, some continuous variation in the phenotypes is observed, usually in a bell-shaped (Gaussian) distribution.

7.3.5 Skin Pigmentation in Humans

Skin Color of Offspring of Two Mulatto Parents

Let us assume that three genes on different (to justify independent assortment; Section 6.2.5 in Chapter 6) human chromosomes control the human pigmentation phenotype (a big and faulty assumption!). Going further, if the dark pigmentation alleles A, B and C are dominant over the light pigmentation alleles a, b and c, then a cross between a very dark person (AABBCC) and a very light person (aabbcc) would result in a mulatto person with a genotype AaBbCc. The progeny of the union of two mulattos using an AaBbCc×AaBbCc cross shown in a Punnett square would show 27 different genotypes and seven different phenotypes (shades of skin color; Figure 7.7). The colors corresponding to the numbers 0 through 6 in Figure 7.7A are shown in Figure 7.7B.

If the color of a person's skin is approximately proportional to the number of dominant alleles (A, B and C), then a curve of the total number of dominant alleles against their frequency (Figure 7.7B) will reflect, ideally, the distribution of shades of skin color in a large population of people—a very rough bell shape! Polygenic control of a phenotype operates in a large number of other human physical characteristics, including

A

Egg gametes	Sperm gametes							
	ABC	**ABc**	**AbC**	**aBC**	**Abc**	**aBc**	**abC**	**abc**
ABC	ABC ABC 6	ABc ABC 5	AbC ABC 5	aBC ABC 5	Abc ABC 4	aBc ABC 4	abC ABC 4	abc ABC 3
ABc	ABC ABc 5	ABc ABc 4	AbC ABc 4	aBC ABc 4	Abc ABc 3	aBc ABc 3	abC ABc 3	abc ABc 2
AbC	ABC AbC 5	ABc AbC 4	AbC AbC 4	aBC AbC 4	Abc AbC 3	aBc AbC 3	abC AbC 3	abc AbC 2
aBC	ABC aBC 5	ABc aBC 4	AbC aBC 4	aBC aBC 4	Abc aBC 3	aBc aBC 3	abC aBC 3	abc aBC 2
Abc	ABC Abc 4	ABc Abc 3	AbC Abc 3	aBC Abc 3	Abc Abc 2	aBc Abc 2	abC Abc 2	abc Abc 1
aBc	ABC aBc 4	ABc aBc 3	AbC aBc 3	aBC aBc 3	Abc aBc 2	aBc aBc 2	abC aBc 2	abc aBc 1
abC	ABC abC 4	ABc abC 3	AbC abC 3	aBC abC 3	Abc abC 2	aBc abC 2	abC abC 2	abc abC 1
abc	ABC abc 3	ABc abc 2	AbC abc 2	aBC abc 2	Abc abc 1	aBc abc 1	abC abc 1	abc abc 0

B

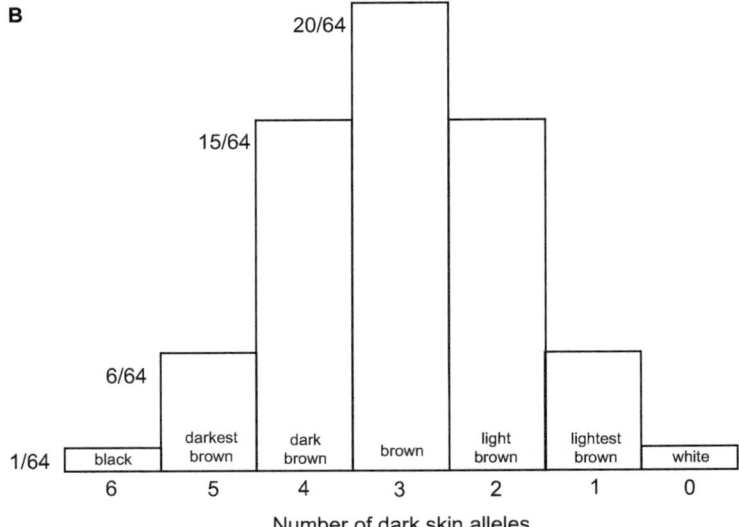

20/64

15/64

6/64

1/64

black	darkest brown	dark brown	brown	light brown	lightest brown	white
6	5	4	3	2	1	0

Number of dark skin alleles

height, weight and eye color. A large number of inherited human diseases are also controlled by polygenes. These include, for example, diabetes, hypertension, peptic ulcers and coronary heart disease (but not cystic fibrosis). This variety of diseases does not, therefore, have a simple inheritance pattern, thus making their genetic analysis and counseling very difficult. Most inherited traits in animals are polygenic. Some examples are: size, speed, milk and egg production, growth rate, numerous inherited diseases, and longevity.

7.3.6 Epistasis

Genes rarely operate in isolation. Interactions with other genes are frequent. One gene may suppress or modify the effect of another non-allelic gene at another locus. A masking gene is said to be epistasis to the subordinate (hypostasis) gene. The concept is well illustrated by a consideration of albinism.

7.3.7 Albinism

Albinism can probably occur in every animal that can produce the color pigment melanin (Section 7.3.9). The inability to form melanin in the eye, skin and hair leads to oculocutaneous albinism, an autosomal recessive inherited condition (Section 6.2.11) with absence of pigmentation in the eyes, skin and hair. The gene responsible for albinism is epistasis to the genes that control the color of a person's hair or skin. Ocular albinism affects only the eyes.

Figure 7.7   (A) Punnett square for the mating of two mulatto parents. The various combinations of the A, B and C, and a b and c sperm and egg gametes are shown. In the Punnett square, the numbers represent the number of capital letters, and therefore the pigmentation shade displayed. For example, the top left corner numbered 6 is an ABCABC genotype and the darkest pigmentation. The square abcabc labeled 0 is the lightest shade. The children could show one of seven shades of skin color. It is assumed that the number of dominant alleles (A, B and C) dictates increasing color with a three-gene control. (B) Gaussian distribution of skin colors in the offspring of mulatto parents. The numbers on the curve indicate the fraction of the population.

7.3.8   Animal Traits

Black, Brown and Albino Mice Coats

Let us consider mice that have black, agouti (brown) or albino coats. Mice have a gene involved in controlling pigment colors. One dominant allele (A) may code for agouti (the wild type of the mouse) and the other (a) may code for a black fur color. Were there no other considerations, we would anticipate brown- and black-colored mice, resulting from various crosses in numbers already described. Another gene, however, has to be considered. This albinism gene (c) may be responsible for the color pigments, even being deposited in the hair shaft. Only if the recessive albinism gene is in the homozygous state (cc), but not with the CC or Cc genotype, will melanin production be blocked. In this case, since the gene is epistasis to the pigment gene, there is no color (albinism) regardless of the state of the A locus, be it AA, Aa or aa. If a pure breeding black-haired mouse, (genotype aaCC) is crossed with a pure breeding albino mouse (genotype AAcc), an agouti mouse (AaCc) results. Dihybrid crossing of two agouti mice produces agouti, albino and black mice in the ratio of 9 : 4 : 3 (Figure 7.8).

Black, Brown and Golden Labrador Dog Coats

Similar considerations apply to the origins of the three coat colors in Labrador dogs, namely black (B gene, dominant), brown (b gene, recessive) and golden, not white (e gene). The golden epistasis gene will again dominate the black and brown genes but only in the homozygous state, and then the Labrador has a golden coat irrespective of the presence of BB, Bb or bb combinations. Otherwise, the black (B) and brown (bb) colors will appear, as shown in Figure 7.9.

Table 7.1 summarizes this behavior for these and two other animals. The ratios of phenotypes seen with the mice and the dog are the same in the other animals.

7.3.9   Pigmentation

Pigmentation is probably one of the most recognizable phenotypes in animals. It has been used in the book as an example of

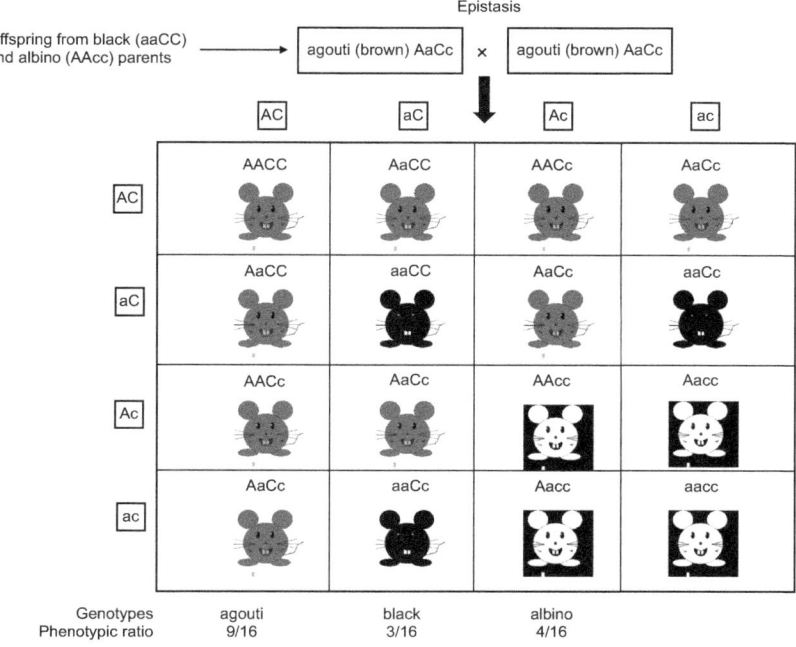

Figure 7.8 Punnett square depiction of the impact of the albino gene on the coat colors of mice. Any mouse with a cc combination will be albino regardless of the state of the A allele. Otherwise, any mouse with an AA or Aa combination will be agouti. Only mice with an aa combination will be black provided the cc combination is not present. The genotypes correspond exactly to those in the dihybrid cross involving guinea pigs (Figure 6.6), with the agouti mouse equivalent to the black, short-haired guinea pig.

Figure 7.9 Black, brown and golden Labrador dogs and their corresponding genotypes.

Table 7.1   Coat colors of several animals.

Animal	Dominant color	Recessive color	Epistasis color
Mouse	Agouti	Black	Albino
Horse	Brown	Tan	Near-white
Guinea pig	Black	Brown	Albino
Labrador	Black	Brown	Golden

autosomal dominance, codominance, partial dominance and recessive inheritance patterns. It also features in X-inactivation, expressivity, pleiotropy, polygenes, epistasis and epigenetics. It thus merits a separate section. Melanin is the major coloring agent in animals. The amount and distribution of eumelanin determines black and brown skin color, and is always the major constituent of epidermal melanin. Pheomelanin is responsible for yellow to reddish coloration in the skin, hair, eyes and feathers in a large variety of animals. High levels are found in the yellow–red hairs of mammals and there are detectable levels in human skin of all colors. Melanocytes are cells located in the basal level of skin. In their cytoplasm, melanosomes have aggregations of the enzyme tyrosinase, which controls the synthesis of either of the melanins from tyrosine and DOPA (Figure 7.10).

In particular, *tyrosinase (TYR), melanocortin 1 receptor (MC1R) and agouti signaling protein (ASIP)* genes play significant roles in the implementation of pigmentation.

TYR *Gene*

The *tyrosinase (TYR)* gene is responsible for the full color in cats. It encodes an enzyme, tyrosinase, which is required in the first step of the production of melanin (Figure 7.10).

Siamese Cats.   This is a slender, short-haired breed of cat with blue eyes, a pale coat, and dark ears, paws and tail tips. It results from a missense mutation (G302R) of the *TYR gene* called *Himalayan*, which encodes a form of the tyrosinase enzyme, whose activity is temperature sensitive. This means that, in the warm body parts of the Siamese cat (and in the newborn kitten, straight from the warm mother's body), the enzyme is

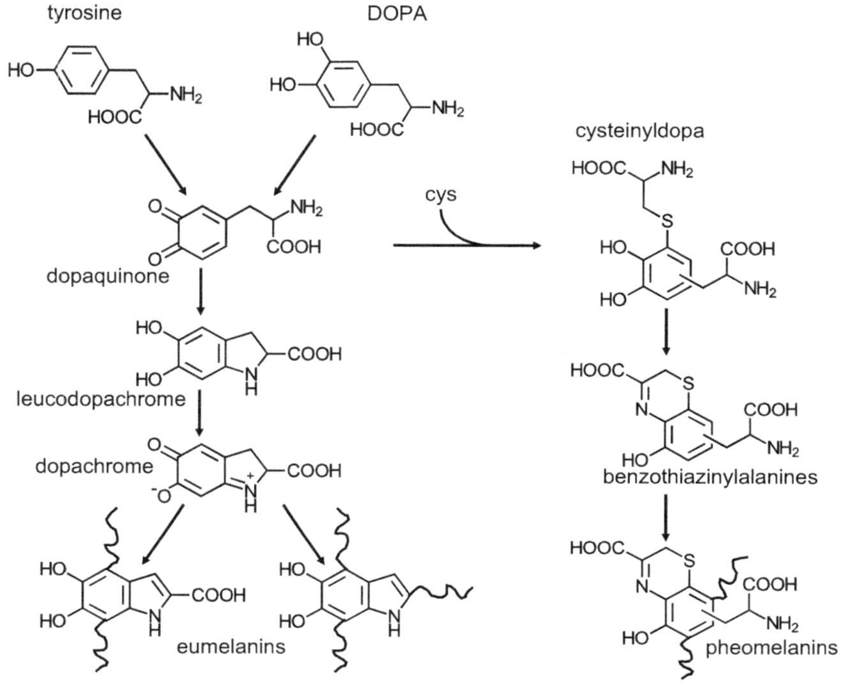

Figure 7.10 The synthesis of melanin from tyrosine and DOPA. The curly lines on eumelanin and pheomelanin indicate points of attachment of many identical molecules to form an extended polymer. The details are uncertain.

ineffective and the color of the animal is off-white, gray or cream. It is not completely white because the *Himalayan* gene is dominant over an albino gene. In the extremities of the cat (the ears, paws, face and tail tips), where the temperatures are lower, the enzyme action kicks in and the characteristic black pointed pattern (which is even darker in the winter) is displayed. This behavior is shown in an experiment in which the application of an ice pack for several days to a region of shaved hair on a Himalayan rabbit results in a black-hair patch growing instead of the original white color. The temperature-sensitive Himalayan mutant also occurs in the mouse, rabbit, guinea pig and other animals. All have light pigmented bodies and somewhat darker extremities, but the coloring will vary within the breed.

The Burmese cat also has a *Himalayan*-type gene, with the mutation G227W. The Burmese cat is darker than the Siamese and has golden eyes. The Tonkinese cat is a recent breed with mink coloring and results from the crossing of the codominant Siamese and Burmese alleles.

Humans. The *TYR* gene has been encountered in oculocutaneous albinism, OCA1 (Section 6.2.11 in Chapter 6). A very rare missense mutation (R422Q) of the gene on human chromosome 11q results in a temperature-sensitive (TS) tyrosinase, which behaves similar to that in animals. A few years after birth, sufferers of OCA1-TS develop darker hair on the cooler areas of the body, such as the arms, feet and chest. On warmer areas, the scalp, for example, hair is generally white. Investigators suggest that the enzyme is rendered inactive at the higher temperatures (37 °C) because it is retained in the endoplasmic reticulum and destroyed there, rather than being moved to the membrane, allowing it to function normally at the lower temperature (31 °C).

MCIR Gene

The *MCIR* gene is at the heart of color variations in a variety of animals.

Rock Pocket Mouse. The melanocytes of the normal (wild type) rock pocket mouse produce more pheomelanin than eumelanin, leading to a sandy-colored mouse. The mutated version of the *MC1R* gene triggers increased production of eumelanin and a dark-coat-colored phenotype results. In the southwest US, wild-type mice occupy light-colored rocks and sand, and are the light-colored type (genotype aa). In the dark rocks of ancient lava flow, the mutant mouse (genotype bb and ab), with a dark color, is encouraged to reside because this eases the problem of their being seen by predators, such as owls. The genetic basis for this adaptive melanism was first elucidated using the behavior of these mice. Once a variation occurs, it becomes the major form in the population and offspring are more likely to survive and reproduce. The behavior described for the rock pocket mice is reminiscent of the story of the peppered moth. Recent work has ascribed the change from the light-colored moth to the black form, encouraged by the darkening of trees in the British industrial revolution, to a mutation of a single gene, called *cortex*.

The mutation involves insertion of multiple copies of a transposon into a genetic intron. Transposons are sequences of DNA that move from one location to another in the genome.

Humans. The *MC1R* gene on the q-arm of human chromosome 16 encodes MC1R protein, which is located on the surface of melanocytes. Many mutants of the gene lead to differences in skin and hair color. When the gene is activated, the protein controls the formation of eumelanin and, when blocked, makes pheomelanin. Two copies of certain mutations, for example, R151C, are common in red-haired people. Possession of two copies of an *MCIR* gene mutation also increases the number of gene mutations associated with melanoma and an increased likelihood of developing skin cancer. Since the MC1R enzyme belongs to a family of receptors that include pain receptors, the gene mutation also appears to influence redheads' sensitivity to pain (in a visit, for example, to the dentist!).

ASIP Gene

All mammals have the *ASIP* gene. The gene is located in the agouti locus on mouse chromosome 2. It contains many alleles and mutations. The *agouti* gene encodes a cysteine-rich protein that directs the melanocytes in the hair follicle when to change from making black pigment to making yellow pigment. Forms that concern us in the discussion of epigenetics (Section 7.4.1) are:

- Agouti solid black or brown, allele a.
- Agouti mottled brown, often referred to as the wild type.
- Agouti lethal yellow (A^Y), which we encountered when lethal alleles were discussed in Section 6.2.4 in Chapter 6. In the heterozygous state, the mouse is a rich yellow or orange color due to pheomelanin, irrespective of the wild-type (agouti) partner allele. The homozygous (A^YA^Y) embryo dies at implantation.

- Agouti viable yellow (A^{VY}) is, again, another dominant mutation (like agouti lethal yellow) over the wild form (a). $A^{VY}A^{VY}$ is again a lethal genotype.

Both forms of yellow mice display pleiotropy since they exhibit a number of different health problems, including obesity, insulin-resistant type II diabetes, tumors and an earlier-than-normal death. The agouti mouse has played an important role in epigenetics.

7.4 EPIGENETICS

The DNA structure is not the "be-all and end-all" of genetic behavior. This statement is a lead in to epigenetics. It is not easy to formulate a hard and fast definition for epigenetics. A useful one is that an epigenetic trait is a stably *heritable phenotype resulting from changes in a chromosome without alteration in the DNA sequence*. The italicized phrase appears in most definitions. A number of changes have been registered. The most examined involve methylation of the DNA [primarily at the $5'$ position of cytosine (C), Figure 1.5, in a CpG island] and histone proteins with which DNA associates, as well as acetylation of histone (Section 1.3 in Chapter 1). Methyl groups are epigenetic markers that induce tight folding of the chromatin (DNA plus histone proteins) and thereby inhibit reading of the gene, *i.e.*, they turn the gene off. Markers, such as acetyl groups, on the other hand, tend to open up the chromatin and spur gene activity. These acetyl groups are generated by acetylation of lysine on amino-terminal tails of core histone proteins. The enzymes responsible for DNA methylation are DNA methyltransferases (DNMTs). Those responsible for DNA demethylation are uncertain. The enzymes responsible for adding and removing acetyl groups are histone acetyltransferase (HAT) and histone deacetylase (HDAC), respectively.

A major difference between genetic and epigenetic effects is that genetic mutations can alter, reduce or increase the amount of encoded protein. Epigenetic changes, in contrast, tend more to alter the gene activity by turning genes on or off, and the effect is well illustrated by considering an agouti mouse.

7.4.1 More on the Agouti Mice

A normal, healthy wild mouse may have a brindled brown pattern, with a hair shaft structure of black at the tip, yellow in the middle and black at the base. The *ASIP* (*agouti*) gene is essential for the middle, lighter band but is switched on and off in a cyclic fashion and mainly silenced during development of the mouse. A mutated *agouti* gene (allele a) may never be switched on and leads to a solid brown or black animal. Another mutated version (allele A^{VY}) has the *agouti* gene permanently switched on, resulting in the agouti viable yellow form. The mouse is totally yellow and has serious health problems. A strain of yellow mouse with the (A^{VY} a) genotype was bred and mice from a single litter examined. Since the A^{VY} allele is dominant over the a allele, the litter would all be expected to be totally yellow since the genotypes of all the kittens are identical. Imagine the original investigators surprise when the litter of mice had a range of colors from a brindled brown mouse through a slightly mottled to a yellow. It is apparent that the basic DNA structure with identical DNA sequences cannot, in this case, explain the different phenotypes. For an explanation, we need to examine the retrotransposon region upstream from the *agouti* gene in the agouti viable yellow mouse (Figure 7.11).

This piece of inserted DNA codes for an abnormal RNA, which messes up transcription of the *agouti* gene and keeps it switched on permanently. This will lead to the yellow mouse. The retrotransposon region contains CpG islands that can be methylated. If it is highly methylated, the abnormal RNA is not activated, the undesirable gene expression is silenced and the normal banded brown coat color returns. If it is not methylated, the agouti yellow form results. Variable methylation, which is possible in the various embryos, leads to the variable coat colors. The epigenetically derived color of one mouse tends to persist to a certain extent in its future generations. This variation in epigenetically derived phenotypes during early development is a partial explanation for certain differences that might not have been expected in identical twins.

Diet Effects

Reinforcing this explanation, if a female agouti viable yellow mouse is fed with mouse chow containing methyl supplements,

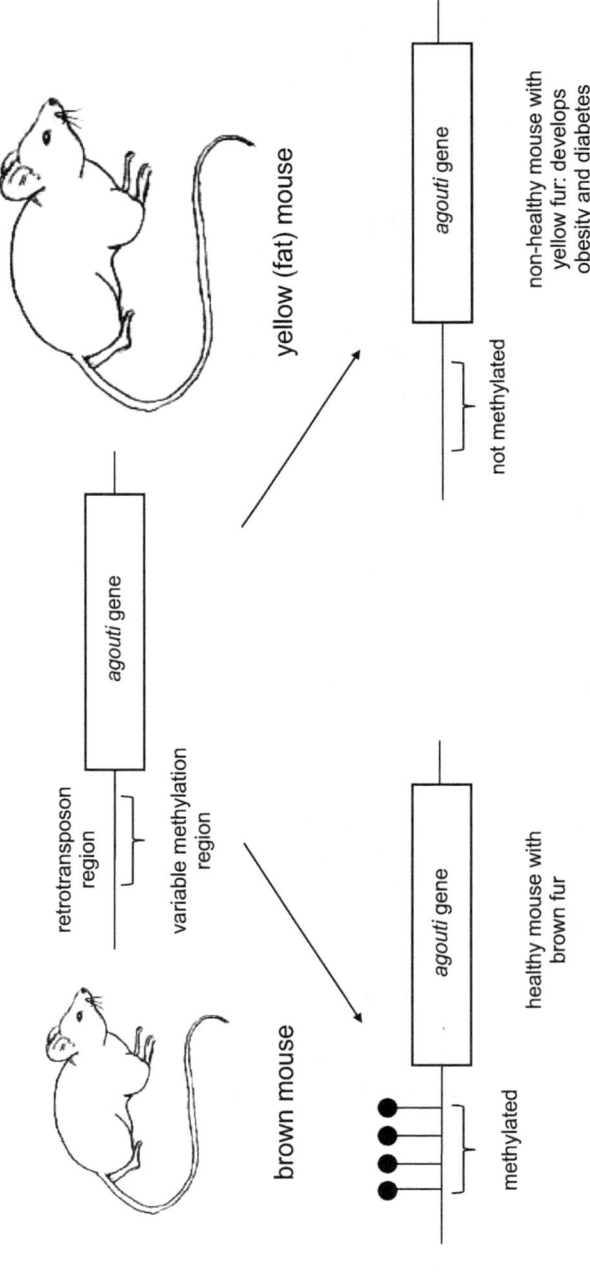

Figure 7.11 The effects of methylation on the variable methylation region upstream from the agouti gene in the agouti viable yellow mouse.

such as vitamin B-12 or folic acid, during pregnancy, brown newborns primarily result, with the gene heavily methylated and silenced. The brown pups stayed healthy for life. Without this dietary supplement, the litter is mainly composed of yellow, unhealthy mice (see Figure 7.11).

7.4.2 Role of Epigenetics in Cellular Processes

Epigenetic factors, such as methylation of CpG dinucleotides, play key roles in many cellular processes. The "correct" DNA methylation pattern is important for successful cell differentiation and embryonic development. The work with the yellow mice tells us that DNA methylation can be important in mediating gene expression. This is a complex process but, in general, high levels of gene expression are associated with low methylation of gene promoters. Most CpG motifs, which are potential DNA methylation targets, are not random in the genome but are often adjacent to the transcription start site, where it impedes transcription. Methylation is important in genomic imprinting (Prader–Willi and Angelman syndromes; see Section 7.4.3) and methylation of histones is important in X chromosome inactivation (Section 7.1.3). The X chromosome that is inactivated in females is silenced by hypermethylation of the histones associated with that chromosome. The particular chromosome inactivated stays methylated and persists from the early embryo until the death of the female. DNA methylation is a long-term conversion preserved during replication and cell division.

7.4.3 Human Disorders

A number of disorders (particularly neurological) have been associated with errors in methylation (epimutations). Genes that are used in defense against, for example, cancer may be silenced by aberrant epigenetic mechanisms.

- DNA methylation plays a large role in genomic imprinting, including uniparental disomy. In Prader–Willi syndrome (Section 7.1.6), there is a lack of expression of the maternally inherited region of chromosome 15 since the maternal allele is hypermethylated and this silences the transcription of genes on the maternal chromosome. A similar situation pertains in Angelman syndrome, except that now it is the paternal allele that is silenced by hypermethylation. The methylation pattern is used clinically to diagnose the diseases.

- The abnormally high (CGG) expansion associated with fragile X syndrome (Section 4.4.1) is only an indirect cause of the gene failure. The expansion triggers hypermethylation, for unknown reasons, of the CG dinucleotides in the triplet repeats and the associated CpG island in the 5′ UTR part of the *FMR1* gene. This silences the gene and results in no encoded protein, which is required for brain development and is normally not methylated.
- The association of epigenetic changes with certain cancers has encouraged the development of epigenetic-directed drugs for their treatment. Focus has been on the DNMT and HDAC enzymes involved in implementing methylation and acetylation (Section 8.5.2 in Chapter 8).

7.4.4 Dutch Hunger Winter

Normally, it is not possible to carry out the types of experiments on humans that we have seen with mice. However, a horrible episode in World War II has allowed some assessment of the effects of food deprivation on a female pregnancy. For a few months from November 1944, a Nazi blockade created famine conditions in Holland. It has been possible to assess the people who resulted from pregnancy during the famine. Those women who were exposed to the famine conditions during the first trimester of their pregnancy had children with normal birth weights but who later experienced obesity and other health problems not shown by their siblings. These were passed on to subsequent generations. Mothers who were in their last trimester of pregnancy during the famine had children with lower than usual weights at birth but remained underweight and did not have the health problems experienced by the other group. The genetic analyses of these two groups proved very interesting. Recall that the imprinted *IGF2* gene plays an important role in growth and development of a fetus (Section 7.1.4). It was found that this gene had a different DNA methylation pattern in an adult who had been exposed to famine in the early period of the mother's pregnancy than unexposed siblings or those who had been exposed in the last trimester of their mothers' pregnancy. It seems that prenatal exposure to food deprivation in the early fetus development can lead to epigenetic changes, which affect a person's health and subsequent generations.

Examples of the Impact of Genetics in Forensics, Agriculture and Medicine

8.1 DNA PROFILING IN HUMANS

About 0.05% of a person's genome varies from that of another person (unless we consider monozygotic twins; see Section 8.3.2). This variation in the DNA sequences is termed polymorphism and is found in the non-gene portion (around 98.5% of a human genome). This variation is used in both nuclear and mitochondrial DNA profiling or fingerprinting. An obvious application is in forensics, where it may be used to establish or eliminate a person's involvement at a crime scene. There have been hundreds of post-conviction exonerations, including a number of persons on death row in the US, by using DNA profiling. A start is being made at forensic DNA phenotyping. An idea of a person's hair, eye and skin color, as well as other physical features, from the DNA left at a crime site may help in giving a clue as to, or ruling out, a possible suspect (see also the examination of Richard III's skeleton in Section 6.3.9 in Chapter 6). It has been used to establish fatherhood more definitively than by the older blood group diagnosis, which can only be used to exclude potential fathers. Adopted people now

Animal Genetics for Chemists
By Ralph G. Wilkins
© Ralph G. Wilkins 2017
Published by the Royal Society of Chemistry, www.rsc.org

have the direct means to confirm a suspected biological parent. These approaches have been mainly aimed at humans but are now beginning to be applied to animals. Tracking the source of an infectious disease and counting the number of endangered animals in a specific area are two such examples.

8.1.1 DNA Analysis in Forensics

The following basic steps are employed for extracting and analyzing the DNA for examination:

- **DNA source.** The DNA must be extracted from the cell nucleus. So, for example, the white (but not red) blood cells, epithelial cells of the salivary and tear ducts (not saliva or tears) or hair follicles (not hair) must be used. A favorite source is the cells from a mouth scrape. Bones have osteocyte cells that contain DNA and, being hardy, are a useful source for very aged samples (Section 1.7.3 in Chapter 1).
- **DNA extraction.** Protease enzymes (protein-destroying) are used to breakdown the proteins of the cell wall and other cellular structures. The DNA is unharmed by this treatment. The DNA is released into solution and chloroform is added, whereupon the DNA in the aqueous layer can be separated from protein in the organic layer. The addition of alcohol to the upper aqueous layer precipitates DNA, which is filtered off.
- **DNA analysis.** A combination of PCR, capillary gel electrophoresis and the examination of STRs are most used. More than one locus on different chromosomes, each containing STRs, can be examined in the one reaction mixture. A pair of primers with specific flank sequences (Section 2.3.2 in Chapter 2) for each locus is used, with one primer labelled with a fluorescent dye. The target DNA, together with the primers, *Taq* polymerase and nucleotide bases, is amplified by PCR (Section 2.3.1 in Chapter 2). Thirty cycles produce around 10^8–10^9 copies of the DNA so that 1 ng (10^{-9} g) will yield 0.1–1.0 g. The DNA in each locus in the resultant product can be separated by capillary gel electrophoresis (Section 2.1.1 in Chapter 2) using the fluorescent label for loci assignment. The resultant electropherogram shows the length of the STR at each locus, and therefore the number of repeats.

8.1.2 How does STR Fingerprinting Work in Practice?

In a particular locus, one person has a variable number of STRs for each of the two human chromosomes derived from the mother and father. Another person will have the *same* type of STR at the *same* location but will probably have a different number of repeats in one or both of the two chromosomes.

Let us consider locus 1 on human chromosome 7, which has STR values ranging from 5 to >14 repeats of a GATA sequence. A person may have the number of repeats for the 1 and 2 alleles arising from the two parents, shown in Table 8.1 for this locus. It is known that 1% of the general population will have the 14 and 3 combination of STRs. If we apply this reasoning to five loci on different chromosomes (Table 8.1), we find that the chances that another person will have the same pattern of STRs will be $0.01 \times 0.03 \times 0.02 \times 0.01 \times 0.02$, *i.e.*, 12 chances out of 10 billion (about one in a billion).

The FBI Combined DNA Index System (CODIS) uses 13 autosomal loci (more are anticipated) for DNA fingerprinting purposes. Each is a tetrameric STR sequence (for example, TCTA) with anything from 8 to >40 repeats. It is easy to see that the likelihood that two persons will have the same STR arrangements in all 13 loci is astronomically low and this is the basis for the assignment of DNA found at a crime scene to a suspect. The European agencies use a different combination of loci for DNA profiling.

Another method used as a basis for DNA profiling in forensics is SNP (Section 4.3.1 in Chapter 4). More than 90% of the genetic variations between humans stem from SNPs. They are also more common in the human genome than STRs. One can, therefore, get results from smaller fragments and are therefore handy when dealing with degraded DNA. This method is unlikely to be used

Table 8.1 STR numbers for the maternal and parental chromosomes in five different loci and the percentage of the general population with those patterns.

Locus	STR number for allele 1	STR number for allele 2	% of population with this combination
1	14	3	1
2	7	11	3
3	2	16	2
4	15	8	1
5	1	13	2

routinely in forensic analysis, however, since the STR system, which is less costly to use, is also well established. In certain circumstances, it may be essential, however. Y-STR (Section 6.3.7 in Chapter 6) in the male Y chromosome shows the same sequential structure as autosomal STR, which it can supplement or even replace.

STR Paternal Testing

STR bands in a child's DNA will result from those of the biological parents alone. If the child has an STR band that does not exist in the mother (which will be about 50%), it must be present in the father, otherwise it indicates that he is not the biological father. The more STR bands in various loci that are examined, the more definite is the result (see also the Siberian Husky in Section 8.1.4).

8.1.3   Mitochondrial DNA

Sometimes, there are limited amounts of nuclear DNA available for analysis, while sufficient mitochondrial DNA might be available since most human cells contain hundreds of copies of mtDNA. In addition, mitochondrial DNA is more stable than nuclear DNA so is useful for examining ancient or highly degraded samples. In the event that a person is not available, any maternally related person may provide a sample (Section 6.4 in Chapter 6). Because of its relatively small size, it is possible to do a complete DNA analysis. It is usually more convenient, however, to carry out DNA profiling by Sanger sequencing or pyrosequencing of the two hypervariable regions (HV1 and HVII) of mitochondrial DNA (Section 1.7.4 in Chapter 1). These are contained in a non-coding D-loop control region, which is about 1000 bp in length. There is a high number of sequence variations (nucleotide polymorphism) in the two hypervariable regions HVI and HVII of mitochondrial DNA (Section 1.7.4 in Chapter 1). DNA variations here differ by as much as 1–3% between non-related individuals. PCR (Section 2.3.1 in Chapter 2) is used to amplify the two regions using flanking primers (Section 2.3.2 in Chapter 2). Both hypervariable regions are then sequenced and a comparison made between two individuals' patterns for forensic analysis. A forensic database of human mtDNA has several thousand sequences available to assess the significance of a match. The possibility of HVI and HVII heteroplasmy (Section 1.7.4 in Chapter 1) should be considered in a forensic analysis.

8.1.4 DNA Profiling in Animals

SNP, minisatellites and microsatellites are found in the DNA of all animals (Section 1.5.1 in Chapter 1). It is, therefore, possible to apply all types of DNA profiling and sequencing to animals. Companies exist that will perform SNP, VNTR and STR analysis on all kinds of animals, ranging from antelopes to zebra fish! Samples of blood, fur, bones or teeth need only be provided. The profiling technique is similar to that already described for humans. Animal DNA profiling is being increasingly used to support criminal investigations.

- These might involve linking a suspect with a dog. The dog may act as a go-between, matching the animal's hair on a rape victim with those recovered from the suspect's car, for example.
- In one case, the mitochondrial DNA in eight cat hairs attached to a curtain wrapped around a homicide victim matched that of the suspect's pet cat Tinker.[†] This was significant since the mtDNA was a rare type for domestic cats. The killer was jailed in 2012 on this and supporting evidence.
- The first determination of canine parentage involved the use of minisatellite loci and a comparison of the DNA generated from a Siberian Husky dam, sire and puppy. Fourteen DNA bands from the puppies' paws, not present in the dam, were found in only one of the potential sires, and this was therefore the true sire.
- DNA profiling can be helpful in tracking the source of an infectious disease outbreak. *Brucella* bacteria cause brucellosis, a devastating infectious disease in many animals, including Elk, bison, sheep, *etc.* Different strains of the bacterial genome have different patterns of tandem repeats of eight oligonucleotides throughout two chromosomes.

[†]Although this was the first time that a cat's DNA had been thus used in the UK, the first example was the case of the murder of a Canadian woman in 1994 by her husband, using his white cat snowball (A. Boyle, NBC News, 14 August 2013).

*H*ypervariable *o*ctameric *o*ligonucleotide *f*ingerprints (HOOF-prints) can therefore be used to identify strains of the disease, trace the source of outbreaks of brucellosis and isolate it.

- There is a not a reliable method for counting the number of endangered species in a specific area. This may be changing as a result of DNA profiling. For example, visual counts of imperial and white-tailed sea eagles in a Kazakhstan nature reserve gave a value three times lower than that obtained using DNA. DNA was extracted from eagle feathers around many roosting sites. Counting the number of different DNA patterns gave an estimate of the number of birds.

8.1.5 Genome-wide Association Studies (GWAS)

With the availability of the vast information about the human genome and the use of next generation sequencing technology (Section 2.1.4 in Chapter 2), it is now possible to map 1000s of SNPs in a human genome concurrently and rapidly. GWAS search for SNPs that occur more or less commonly in persons with a particular disease or characteristic compared with those without that particular trait in a large, general population, *e.g.*, blood pressure. People with heart disease, for example, may have a significantly higher percentage of C than T nucleotides at a particular position in the genome. This difference may not directly cause the disease but is an excellent pointer to the region where the disease-causing problem lies. Thousands of specific areas of the genome associated with various diseases and traits have been identified, and a catalog of data extracted from the literature and assessed by experts is available. This information may explain how diseases develop and might be treated. It is hoped that, in time, this will lead to new treatment for diseases and other conditions. For example, GWAS found that age-related macular degeneration was associated with a gene involved in regulating inflammation so that anti-inflammatory therapies might be an (unexpected) option for treating the disorder.

8.2 GENETIC ENGINEERING

Genetic engineering is an all-embracing term covering the laboratory or commercial techniques used to clone (amplify) and manipulate DNA. Cloning is the duplication of a gene, a single cell or a whole animal. It is also variously termed gene, DNA or molecular cloning, or recombinant DNA technology. Gene cloning is implemented *in vivo* using vectors carrying the gene into a replicating cell (Section 2.2 in Chapter 2) or *in vitro* through enzymatic amplification using the PCR technique (Section 2.3 in Chapter 2).

8.3   TINKERING WITH ANIMAL GENES: CLONED, TRANSGENIC AND CHIMERIC ANIMALS

- A cloned animal is, for all intents and purposes, genetically very similar to the animal from which it was derived (see, however, epigenetic modifications in Section 7.4 in Chapter 7). Animals may bear slight phenotype differences, since the environment and other factors play a role.
- In a transgenic animal, a new gene is introduced or an existing gene modified during the cloning of a wild-type animal with the intention of modifying or enhancing a desired phenotype. Alternatively, a faulty gene in an animal with a disorder may be corrected and cloned in an attempt to regenerate an animal without the undesired trait.
- Chimeric animals are derived from different strains of an animal or even from two different animal species.

8.3.1   Cloned Animals

Here, we are concerned with the production of identical copies of an original cell or animal without gene alteration.

8.3.2   Twins

Natural cloning in animals (and humans) is observed in the occurrence of identical (monozygotic) twins, most commonly in cattle, cats, ferrets and deer. Nine-banded armadillos regularly have identical quadruplets! The fertilized egg splits into two separate embryos anywhere between the two-cell stage or as late as the blastocyst stage. The split gives two genetically identical embryos, leading to identical monozygotic (MZ) twins sharing the same placenta (Figure 8.1).

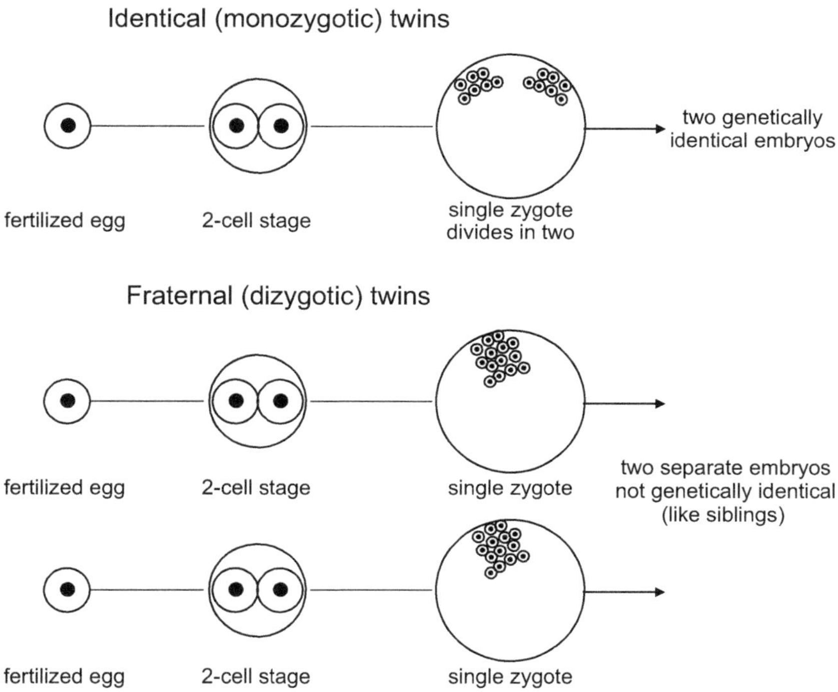

Identical (monozygotic) twins

fertilized egg   2-cell stage   single zygote divides in two   two genetically identical embryos

Fraternal (dizygotic) twins

fertilized egg   2-cell stage   single zygote   two separate embryos not genetically identical (like siblings)

fertilized egg   2-cell stage   single zygote

Figure 8.1   The generation of identical and fraternal twins.

The DNA sequences will almost certainly be identical in such twins and different from either parent, but the two may have different DNA methylation patterns (Section 7.4 in Chapter 7). MZ twins should be distinguished from fraternal, dizygotic (DZ) twins, where two eggs, rather than the one egg usually produced from ovaries, are fertilized to yield two separate zygotes (Figure 8.1). Since fraternal twins begin from a different egg, they are as different (physically, DNA content, *etc.*) as siblings.

8.3.3   The Creation of Cloned Animals: Somatic Cell Nuclear Transfer

Reproductive (animal) cloning is commonly implemented by a procedure termed somatic cell nuclear transfer (SCNT), which is often shortened to nuclear transfer. The technique is illustrated in Figure 8.2 for the creation of a cloned lamb (Dolly).

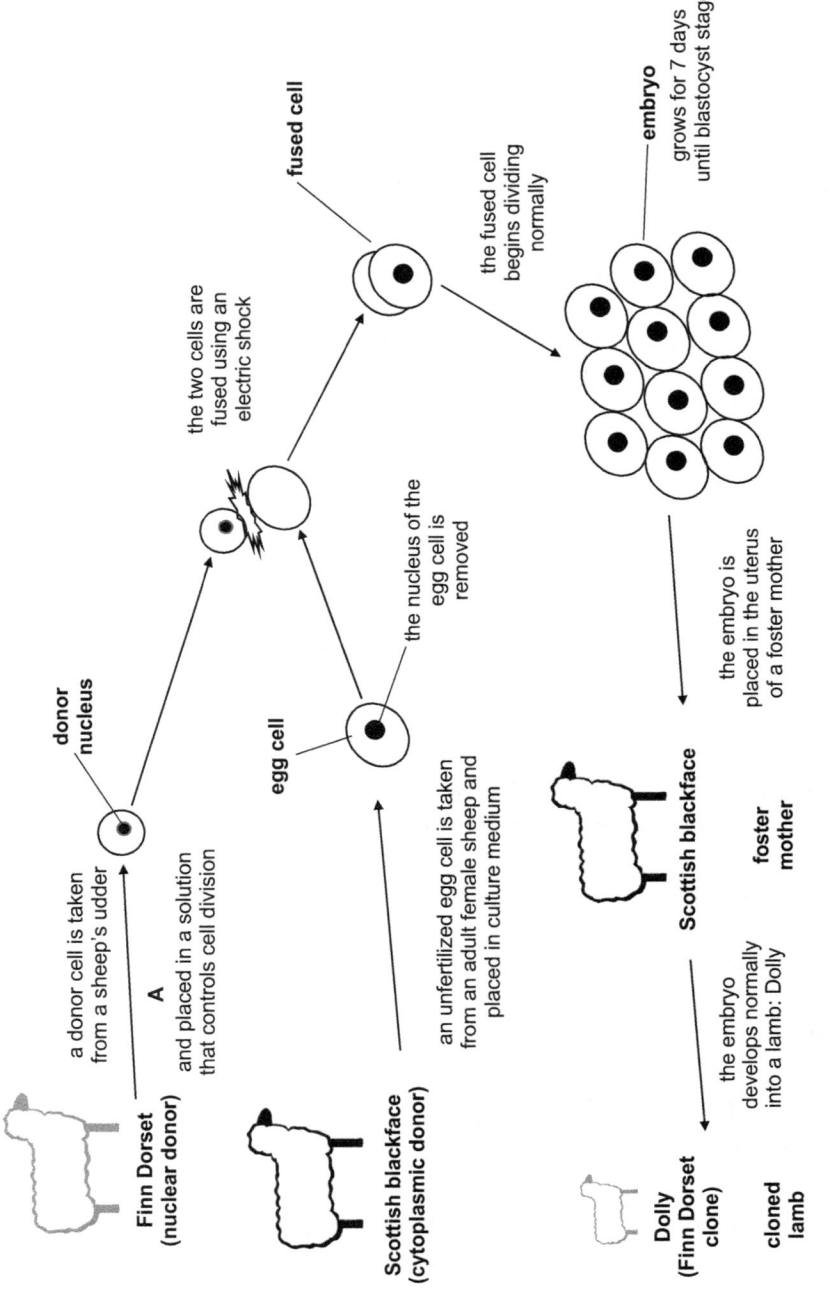

Figure 8.2 Creation of a cloned sheep (Dolly) using nuclear transfer. The donor, a white-faced Finn Dorset ewe, has identical nuclear DNA to that of Dolly. It is distinct from that of the cytoplasmic donor and surrogate mother, a Scottish blackface ewe. [For the production of transgenic animals, the transgene is introduced into the animal's skin cells and grown in the laboratory (position A in the figure) before the donor nucleus is removed.]

No sperm cells are used directly during the procedure. The Finn Dorset sheep that is to be cloned (genetic donor) provides the nucleus from the udder cell of the sheep. A Scottish blackface ewe provides the unfertilized egg cell, from which the nucleus has been removed (oocyte enucleation), to yield a cytoplast (a cell containing only cytoplasm), which contains the cellular machinery necessary to form an embryo. The udder cell is provided with a low concentration of nutrients, which keeps it alive but stops it from growing, and then it is placed near the enucleated egg cell in a laboratory dish and fused by electric shock. This probably punches holes in the egg membrane and allows the nucleus of the udder cell to slip into the egg. (In a variation, the nucleus is removed from the somatic cell and injected into the empty cytoplast egg.) Another electric shock causes the fused cell to begin dividing and develop into an embryo. Cellular division continues and a blastocyst forms in the embryo after several days. The blastocyst, containing 100–200 cells, is transferred to a recipient female surrogate mother (a Scottish blackface ewe ready for pregnancy) after about seven days. The surrogate mother gives birth to an animal after 148 days, which is essentially identical to the Finn Dorset white sheep donor. Dolly, however, contained nDNA from the udder cell, but the animal's mtDNA originated solely from the egg, *i.e.*, it was a chimera with DNA of different origins. It should be emphasized that almost 300 nuclear transfer attempts were necessary before Dolly was obtained. Dolly was not healthy and died at the age of six, half the lifespan of a normal sheep. However, since this pioneering work, four sheep cloned from the same cell line as Dolly have turned nine years of age and are as healthy as naturally born sheep of the same age.

Nuclear transfer is the most common and efficient method for animal cloning. Since the first animal was cloned in 1996, there have been a number of different animals cloned, some 1000 times, by SCNT, ranging from cattle to camels! A commercial cloning service has cloned many dogs at around $100 000 per dog, including pets of wealthy owners and excellent sniffer and rescue dogs. The process has helped boost the population of endangered animals, such as the African wildcat or the Asian wild ox, or to clone animals that have only recently disappeared.

Bucardo Goat

The Bucardo or Pyrenean ibex (wild goat) became extinct in 2000 but living cells from the ear of the last animal, Celia, were frozen in liquid nitrogen. Cloning was carried out in much the same way as with Dolly by nuclear transfer. A nucleus from Celia's preserved cells was injected into an enucleated wild-goat egg. This was implanted into a surrogate Spanish wild and domestic hybrid goat. One of many attempts became pregnant and carried to term. Celia's clone died, however, shortly after birth with a lung defect. It is likely that the attempt at successful cloning will be carried out with an improved technique.

Australian Gastric-brooding Frog

A nuclear transfer approach has been started using freezer-preserved cells from the tissue of this frog, which only became extinct in the 1980s.

The female frog incubates embryos in her stomach by controlling her stomach acidity. Understanding this process by having a number of cloned frogs available may help in the treatment of human stomach ulcers since the creature can halt its production of acid.

Therapeutic Cloning

Reproductive cloning of humans is forbidden worldwide, but at the blastocyst stage of SCNT another route may be taken. In therapeutic cloning the embryo at the blastocyst stage is destroyed to allow the harvesting of embryonic (pluripotent) stem cells from the ICM of the blastocyst. These are grown in a dish containing nutrients and can be induced by known methods to differentiate into specific cell types. Therapeutic cloning might be used to replace injured or diseased tissue, *e.g.*, myocardial cells grown from embryonic stem (ES) cells might be used in the treatment of heart muscle disease. Since the material used, say skin, originates from the patient, the transplanted cells are unlikely to be rejected when transferred back to the patient. It has been shown to be a valid strategy in animal experiments, but its application to humans is in its infancy and has been confined to certain countries outside the US.

Handmade Cloning

A few animals (sheep, pig, horse and cattle) have been cloned using handmade cloning. This approach has been around for some years but has recently been refined. Based on SCNT, it is a simplified alternative, needing less sophisticated equipment and skilled expertise, and is cheaper than traditional SCNT. It is applicable to both cloned and transgenic animals, and comparable birth rates are claimed.

8.3.4 Transgenic Animals

Transgenic animals have a new gene introduced, or an existing gene modified, before their cloning. A specific property may be modified or enhanced (in a wild-type animal) or removed (from an animal with an undesired trait). If the new genetic information is introduced into the zygote with one cell or an embryo containing very few cells, then that information will end up in virtually every cell of the developing animal. Otherwise, mosaicism will result (Section 5.3.4 in Chapter 5). Transgenic, but not cloned, animals increase the genetic diversity of animals.

8.3.5 The Creation of Transgenic Animals

Three methods are available for the creation of transgenic animals, although the embryonic stem cell method has only been successful for producing transgenic mice.

8.3.6 Nuclear Transfer

This method, which is used for producing cloned animals, has been adapted to generate a number of transgenic livestock animals, as well as rabbits and mice. It is favored for the production of bovine transgenic animals. The technique shown for cloning animals (Section 8.3.3) is only genetically modified at the beginning by introducing a DNA construct. This consists of (a) the gene it is desired to introduce, plus (b) a promoter and enhancer so that the gene can be expressed by the host cell, and (c) a terminator stop sequence ensuring that only the gene of interest is controlled by the promoter. A vector may also be required for insertion of the DNA into the host cell. Any genetic modifications are carried out in cultured cells before the nuclear transfer into the enucleated egg (see Figure 8.2).

8.3.7 Pronuclear Injection

This was one of the first techniques used (in mice) and is the more commonly used method. The whole approach is well illustrated by the creation of transgenic goats that are able to synthesize recombinant human antithrombin (rhAT) in their milk to be used by patients with hereditary antithrombin deficiency. The gene construct consists of (a) the milk-specific promoter, which is the regulatory sequence of the goat gene that is only expressed in the goat's mammary cell. This is fused to (b) the target DNA, which is the coding sequence of the human antithrombin gene. This gene construct is microinjected with a very fine needle into a pronucleus (Figure 5.7) of a fertilized egg within hours of fertilization at the one-cell stage, so that all the cells resulting, including the germline, will contain the inserted gene. Fusion of the pronuclei generates the zygote nucleus. The zygote divides to form the two-cell embryo by mitosis. This entity is implanted in the uterus of a *pseudo*-pregnant (foster) mother. (She has been mated with a vasectomized male, so as to stimulate necessary hormone changes in the female.) Mating two heterozygous offspring may produce a homozygote animal. Mating homozygous animals eventually sets up a small herd with the transgenic strain. The goat has now been duped into making human antithrombin in her milk, from which the target protein may be isolated. The method does have the disadvantage of a low integration rate (1–3%) and DNA insertion is random, rather than at a specific site. Nevertheless, it is estimated that one transgenic goat can produce in a year the same amount of antithrombin as 90 000 blood donors. Transgenic goats have also been generated by the nuclear transfer technique, where the majority of the offspring are transgenic.

Hereditary Antithrombin Deficiency Types 1 and 2

The antithrombin (*serpin peptidase inhibitor, SERPING1*) gene on human chromosome 1q encodes antithrombin protein found in the bloodstream. It is an anti-coagulant and is a principal in-hibitor of thrombin, a component of blood coagulation factor IX, which promotes blood clotting. Over 200 mutations of the gene, mostly involving a single amino acid change, lead to the owner having reduced antithrombin (type 1) or an altered antithrombin (type 2). Both lead to the autosomal dominant disease with in-creased risk of developing abnormal blood clots. Antithrombin, commercially marketed as ATryn, has FDA and European Medicines Agency (EMA) approval. The FDA is a US Federal Agency that protects the public against impure and unsafe foods, drugs and cosmetics. The EMA serves a similar role in the approval of certain drugs for people and animals in the EU. ATryn is used particularly for patients with hereditary antithrombin deficiency undergoing surgery and who need large doses of antithrombin to counteract an increased risk of developing blood clots.

8.3.8 Embryonic Stem (ES) Cells

This is quite different from the other two methods (Figure 8.3) and is effective only for mice.

ES cells are harvested from the ICM of a mouse blastocyst. These can lead to all the cells of the animal (Section 1.1.2 in Chapter 1). The ES cells grown in a tissue culture are exposed to the desired gene construct so that some will be incorporated. This will be by homologous recombination (Section 5.1.1 in Chapter 5) with the DNA of the gene having a similar sequence to

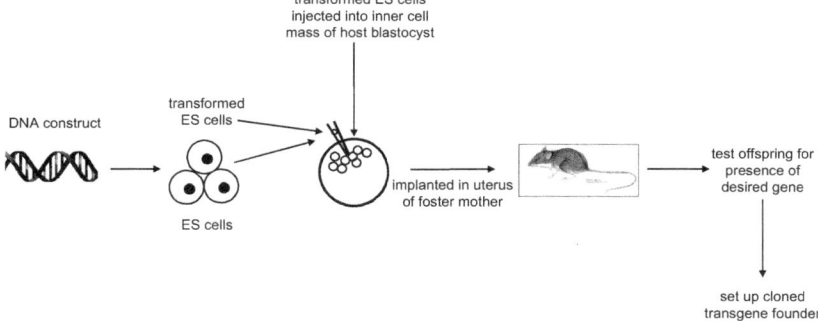

Figure 8.3 Altered embryonic stem cells derived from the mouse blastocyst to create transgenic mice.

part of the mouse genome. The incorporation might be *via* electroporation or microinjection. The very few transformed ES cells resulting are injected into the ICM of the mouse blastocyst. The embryo is then implanted into the foster-mother's uterus and the resulting offspring used to set up a cloned transgene founder by animal cross breeding. Transgenic animals are used in biomedical research and to produce nutritional supplements and pharmaceuticals. Research in the area has been boosted by FDA approval for products from modified clones of chickens, rabbits, cows, pigs and goats.

Recombinant Pharmaceutical Proteins

Bacteria, mammalian cells, transgenic plants and transgenic animals are bioreactor systems that are able to, or can potentially, produce pharmaceutical products. Recombinant therapeutics have been traditionally grown using bacteria (Section 2.2.1 in Chapter 2) or mammalian cell cultures.

- Bacteria are limited in their ability to carry out the post-translational modifications often necessary to produce the final viable product (Section 3.1.6 in Chapter 3). They do, however, grow rapidly and are easy to maintain.
- Mammalian cells do not have the problem of final protein assembly. At present, most products are from bacterial and mammalian cell cultures.
- A number of different transgenic animals have been used, focusing on milk as the protein source. Using the transgenic animal as the bioreactor can produce more complex proteins more efficiently and economically than other methods. It has been estimated that establishing a herd of transgenic animals may cost 1/10th of that from building a commercial cell-culture facility. In spite of this, only three proteins from transgenic animals have reached the drug market in the past 20 years, although several clinical trials are in progress. The drugs sanctioned are ATryn for the treatment of hereditary antithrombin deficiency (Section 8.3.7), Ruconest for the treatment of hereditary angioedema and Kanuma for the treatment of Wolman disease, a very rare lysosomal acid lipase deficiency disease, which prevents the body from

Table 8.2 Transgenic animals and their value.

Animal	Use of transgenic animal
Cow	High quality milk and top-notch burgers obtainable; used to produce human insulin.
Goat	High milk production and short (18 month) generation time (Section 4.5 in Chapter 4); used to produce human antithrombin in milk; both drug and genetically engineered goat have FDA approval.
Mouse	Large variety of genetically identical mice are available for research; healthy and viable mice now produced from human stem cells.
Pig	Source of insulin for type 1 diabetes and coagulation drugs for hemophiliacs; kidney and heart donation; leaner meat.
Rabbit	Used to produce human C1 inhibitor to treat hereditary angioedema.
Sheep	Source of factor IX for hemophiliacs; extra woolly sheep.
Chicken	Used to produce Kanuma in the whites of eggs for the treatment of Wolman disease.

breaking down cellular fatty acids and, in one form, is fatal in infants. Safety and ethical questions have been raised. More than 90% of cloning attempts fail and most cloned animals have poor health and short lives. In agriculture, improved milk production and composition, enhanced prolificacy and reproductive performance, and increased disease resistance in animals are possible. Examples of some transgenic animals, together with their value (actual or potential), are shown in Table 8.2.

Hereditary Angioedema (HAE) Types 1 and 2

The *SERPING1* gene on human chromosome 11q encodes the C1 esterase inhibitor protein, which regulates several inflammatory pathways in the body by inhibiting proteases that are part of the immune system. More than 250 gene mutations lead to hereditary angioedema types 1 and 2. Mutations that cause type 1 occur throughout the gene and lead to reduced amounts of inhibitor in the blood. Mutations in exon 8 of the gene result in type 2 and an inhibitor that functions abnormally. Both lead to the disease, which involves severe swelling in a number of areas in the body, including the airways, which can be life-threatening. Ruconest has been approved for the treatment of acute attacks of HAE in adults. Ruconest is an injected human recombinant C1 esterase inhibitor purified from the milk of genetically modified transgenic rabbits. It is intended to restore the level of inhibitor in the patient's plasma. One value of this and similar animal-derived treatments is that the risk of transmission of plasma-related human infections is removed.

8.3.9 Genetically Engineered Mosquitos

In the examples described so far, transgenic animals have been bred for the benefits of mankind without the deliberate harming of the animal. Such is not the case with the genetically engineered mosquito! The subject has raised a good deal of controversy with pros and cons on both sides. Mosquitos are involved in a number of diseases with many deaths every year. Dengue fever causes over 20 000 deaths each year and 50–100 million infections, with severe illness often for weeks. The *Aedes aegypti* mosquito is a worldwide carrier of a number of viruses, including dengue, chikungunya and yellow fever, which can be spread to humans through a single bite. Drugs to combat these diseases and insecticides directed at the insect have had increasingly limited value as drug- and pesticide-resistant mosquitos have appeared. Can a genetic engineering approach come to the rescue?

One strategy of several is to create a sterile mosquito genetically. A transgenic *Aedes aegypti* has been created containing a gene encoding a lethal protein, which disrupts the cell machinery. This insect survives in the laboratory on a diet containing the antibiotic tetracycline, which suppresses the action of the lethal protein. The females are destroyed and the modified males, which don't bite, are released in the wild and can still mate with a wild-type female mosquito. Eggs hatch normally, but offspring inherit the lethal gene and, without the antibiotic, they die before they can transmit the disease or reproduce. In a limited area in Brazil, the number of mosquitos that spread dengue fever has been reduced by more than 90%. Provisional approval by the FDA has been obtained to use the altered mosquito for field trials in Florida. Areas in the state have seen incidences of Zika fever, which is caused by a mosquito-borne virus, similar to that causing dengue. Because the lethal genetic effect persists for only one generation, the mutant mosquito must be released repeatedly, but this is a safeguard feature. CRISPR (Section 8.4.9) has been used to insert a new gene into an

Indian mosquito that codes for anti-malarial antibodies to combat the malaria parasite. Inheritance of the new DNA is 100% for at least three generations. It is hoped that, ultimately, the insect will be unable to transmit the disease to humans.

8.3.10 Chimera Animals

Chimera animals have more than one genetically distinct population of cells resulting from the fusion of (usually) two fertilized zygotes or early embryos. These may be derived from different strains of an animal or even two different animal species. The resultant animal contains cells from each of the two animals and has two different sets of DNA. A "geep" results, for example, from combining a goat and a sheep embryo, and the animal has wool and hair patches. The phenomenon should not be confused with mosaicism, where the two types of cell arise from one zygote. Mosaicism is relatively common, after all, about half of the world's population is female (see X-inactivation in Section 7.1.3 in Chapter 7). In contrast, the natural occurrence of chimera in animals is rare.

8.3.11 Creation of Chimera Animals: Whole Embryo Aggregation

Several techniques are available for the formation of chimera animals. The simplest one (in principle!), termed whole embryo aggregation, uses an embryo at an early stage (say, containing four cells) that has been removed from each of two or more pregnant animals and the cells gently pushed together in a culture dish. With luck, they will fuse together to form a single embryo (Figure 8.4).

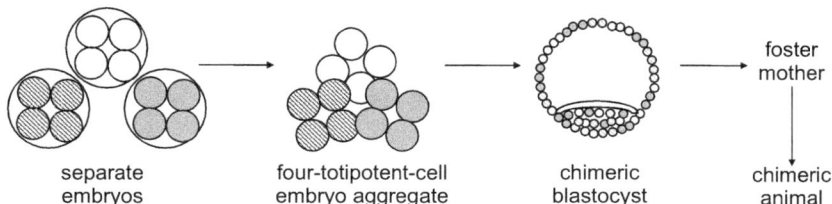

separate embryos · four-totipotent-cell embryo aggregate · chimeric blastocyst · foster mother · chimeric animal

Figure 8.4 Whole embryo aggregation for the formation of chimera animals. Three animals are involved.

If successful, within a few days they will have grown into an early-stage embryo (blastocyst), which is then implanted into a foster mother, who has been mated with a sterile male in order to prepare the uterus. She thus produces a chimera animal, whose eyes, ears, *etc.* all comprise some cells originating from two or more different pairs of parents. In this way, a pair of pure (inbred) mice with black hair and a pure pair with white fur, when treated by whole embryo aggregation, may lead to a chimera mouse with black and white patches. This is different from mating a pure black mouse (dominant) with a pure white one, when all the cells would be of the same genotype and the coat of the offspring would be uniform black. The technique has been used successfully with a number of animals, including the first production of a primate: a chimera rhesus monkey. With mice and other animals, either totipotent (as described above) or pluripotent stem cells from two different animal types can be combined to create an embryo that later becomes a chimera mouse. For some unknown reason (as yet), chimera monkeys can only develop from totipotent stem cells and not from pluripotent stem cells, and only whole embryo aggregation (of several possible techniques) is effective.

When a chimera animal occurs, it probably arises from the fusion of two early embryos into one. As with the case of mosaicism, it is likely to go undetected since, in general, no health problems arise with chimera animals, except chimera cattle, which are much less rare. With these animals, each fetus of a twin is seeded with stem cells from its mate *via* the exchange of blood between the two fetuses. When one fetus is a male and the other fetus is a female, the female twin is "masculinized" from exposure to hormones from the male twin. Such female cattle are called freemartins and have genital abnormalities and are sterile. Chimeras were produced for the first time over 50 years ago in mice and have become a valuable research tool in understanding, for example, embryonic development. Human–animal chimeras, prepared from human stem cells introduced into developing animal embryos, are being used under strict guidelines to study disease, test new drugs and (possibly) be a source of organ transplants.

8.3.12   Humans

Reports are very rare of chimerism in humans, but the phenomenon may go undetected since it is not often recognizable physically. A few cases have been reported recently since DNA diagnoses have been more readily available. Two fertilized eggs in the womb yield two zygotes and, subsequently, fraternal twins (Figure 8.1). If, however, the two zygotes fuse (a very rare event), only one baby results and this baby, termed tetragametic, will have two distinct sets of DNA: its own and its former fraternal twin. These two sets will differ as much as with siblings.

8.4   TINKERING WITH HUMAN GENES

8.4.1   Gene Therapy

Gene therapy is repairing a faulty gene that might lead to a disorder in any location within an animal genome by replacing it with the healthy version.

Recently, the replacement of *the whole complement of genes* in the mitochondria, both faulty and healthy, in humans by a healthy set has been sanctioned in the UK.

8.4.2   Getting Rid of "Faulty" Mitochondria!

A number of inherited, severely debilitating or lethal diseases are associated with mitochondrial mutations. Would it not be very valuable to be able to replace any "faulty' mitochondria containing crippling mutant genes with a "healthy" one? Swapping affected mtDNA with healthy mtDNA in a female would allow her to become a mother with children not having her crippling disorder, which she possibly would have passed on. Such an exchange has been feasible for some time and has been tested on mice, monkeys and humans. It has now been legalized for humans in the UK, with strict safeguards. Two approaches to repair mtDNA are envisaged (Figures 8.5 and 8.6).

8.4.3   Maternal Spindle Transfer (Egg Repair)

A woman's faulty egg is repaired before conception. The spindle apparatus containing the nucleus is removed from the mother's egg, leaving behind the faulty mitochondria. This spindle is then inserted into a healthy donor's egg, from which the spindle has been removed but in which the healthy mitochondria (plus the cytoplasm and membrane) are retained. The result is a

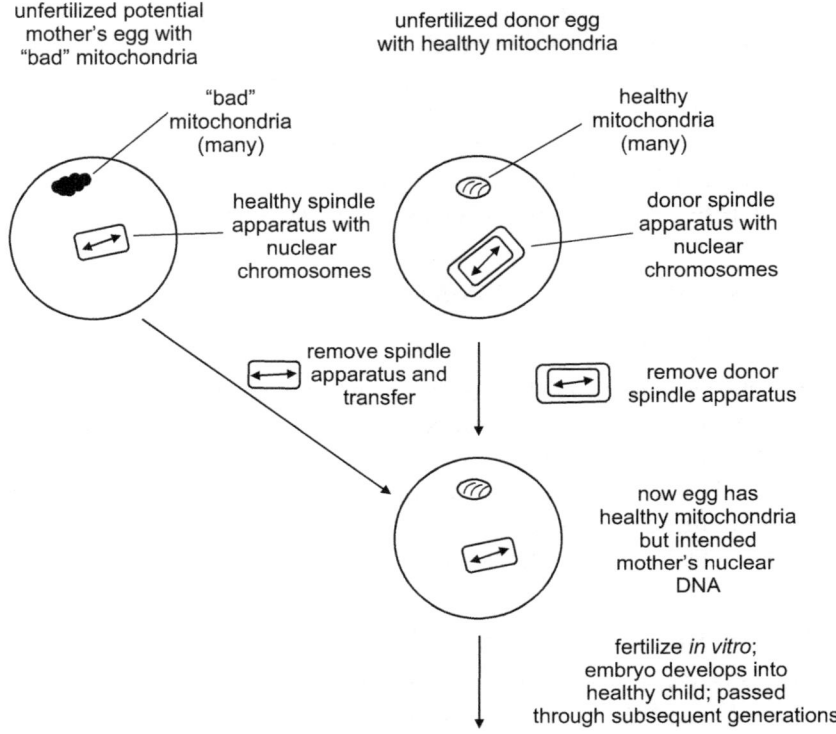

Figure 8.5   Mitochondrial repair using maternal spindle transfer. The spindle apparatus (see mitosis) contains the nuclear chromosomes.

reconstituted egg containing nuclear DNA from the mother and healthy mitochondria from the donor. This can be fertilized with the father's sperm in a clinic and then implanted in the womb and brought to term unaffected by any inherited mitochondrial disease (Figure 8.5).

8.4.4   Pronuclear Transfer (Embryo Repair)

In this approach, the faulty egg is repaired after conception. Pronuclei (Section 5.3.3 in Chapter 5) are removed from the mother's newly fertilized egg, leaving behind the faulty mitochondria. These pronuclei are inserted into a surplus fertilized egg (from a donor couple's IVF cycle), from which the two pronuclei have been removed but the healthy mitochondrial DNA

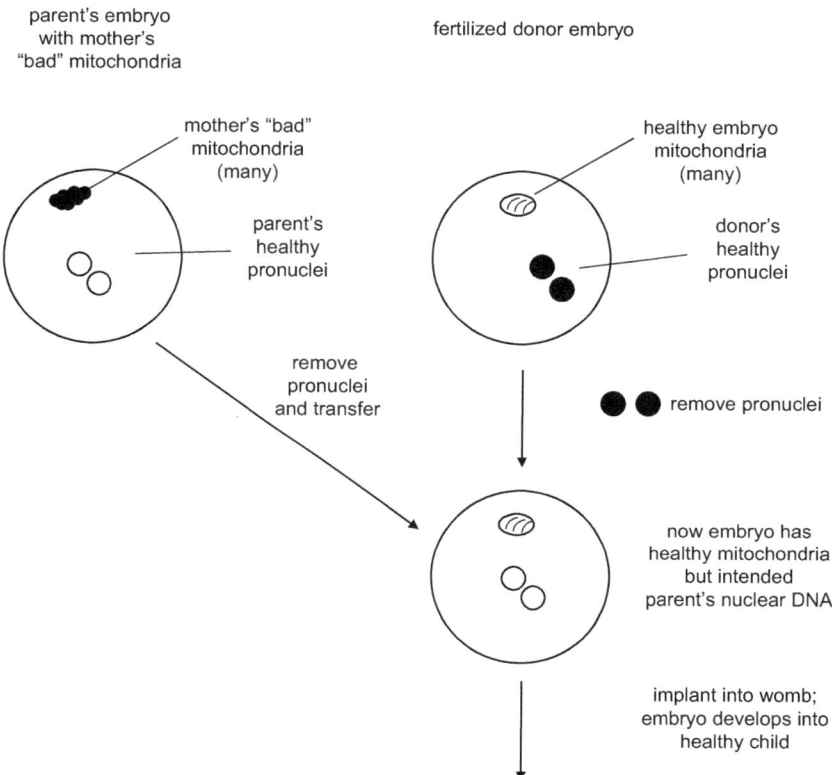

parent's embryo
with mother's
"bad" mitochondria

fertilized donor embryo

mother's "bad"
mitochondria
(many)

parent's
healthy
pronuclei

healthy embryo
mitochondria
(many)

donor's
healthy
pronuclei

remove
pronuclei
and transfer

remove pronuclei

now embryo has
healthy mitochondria
but intended
parent's nuclear DNA

implant into womb;
embryo develops into
healthy child

Figure 8.6 Mitochondrial repair using pronuclear transfer.

has been retained. The repaired egg is implanted in the womb and brought to term (Figure 8.6).

A baby born using one of these two procedures will thus have nuclear DNA from the two parents and mitochondrial DNA from the donor. The donor is theoretically a third parent but her contribution is only, albeit a vital, 2% of the baby's DNA. So far, no three-parent baby has been born.

Use of a third parent and problems arising from possible transfer of some faulty mitochondrion with the mother's nucleus (which has been observed) may be avoided if experiments so far confined to the laboratory can reach clinical fruition. This may be years away but could be very useful in basic research in the intervening years. The around 1 in 4000 persons in the US who have a mitochondrial-based disease usually have a mixture of healthy and mutated mitochondria in every cell. The ratio may define the severity of the disease. Skin cells from such persons can be used to generate a mixture of healthy and disease-induced pluripotent stem cells (Section 1.1.2 in Chapter 1). The disease-free stem cells can be grown and isolated. (Patients with *only* diseased mitochondria cells must be treated differently.) Somatic cell nuclear transfer, as used in the creation of Dolly (Section 8.3.3), is employed. Skin cells have their nuclei removed and injected into a donor's egg cell with its nucleus removed but containing healthy mitochondria. The new egg is used to generate healthy pluripotent stem cells. In both cases, the idea then is to transplant the disease-free stem cells back into the body to generate the required brain, muscle or heart cells with totally healthy mitochondria.

8.4.5   Somatic Cell Gene Therapy

Somatic cell gene therapy is currently attracting much attention. The methodology is effective, albeit difficult. The technique is directed at non-sex cells and only the person receiving this therapy will benefit. This raises fewer ethical issues than those that arise in germline gene therapy, where the genes would be inserted into the reproductive cells/tissues of the human and the alteration would be passed on to the next generation. There are various strategies used in somatic cell gene therapy to make the faulty gene correction. The mutant gene may be reverse mutated back to the normal gene by selective reverse mutation (the most direct approach). The mutant gene may be "knocked out", silenced or, in the most popular approach, replaced with a correct version, which builds the desired RNA and protein. Most commonly, a vector (viral or non-viral) is used to deliver the therapeutic gene to the required cells. In general, viruses are more effective as vectors.

- Adeno-associated virus (AAV) is a small virus with a 4.7 kb genome, which infects humans but does not cause disease, and therefore evokes no immune response. It infects both dividing and non-dividing cells, which is important because brain and heart organs have largely non-dividing cells. AAV comes in a variety of serotypes, which can integrate their genome into various cell types preferentially, *i.e.*, AAV1 and

2 (muscle), AAV8 (liver) and AAV9 (heart and brain). AAV2 is probably the most used and has proved to be a promising vector for treating some inherited rare eye diseases, such as Leber congenital amaurosis and choroideremia—a simple one-gene defect causing destruction of light-sensing cells in the retina. It also holds promise for treating more common age-related eye disorders where the patient is on the verge of losing sight.

* The stripped down HIV virus is proving to be a very useful vector. It evades an immune response and does not disturb cancer-promoting genes (oncogenes). Unlike AAV, it is very effective for installing multiple or chunky genes.

There are two ways to approach the replacement, either *in vivo* by direct vector delivery of the therapeutic gene into the affected tissue in the body or *ex vivo*, with cells taken from the body, genetic alterations made in culture and then transplanted back into the body, Figure 8.7.

8.4.6 Direct Delivery (*In Vivo*)

Direct delivery is the simpler method but is less controllable and efficient than the more complex, indirect cell-based method, where cells can be more precisely manipulated.

Therapeutic genes are isolated from a heathy person and cloned. They are either injected as such or, more usually, packaged in a vector and delivered to a patient's affected body tissue. The whole conglomerate enters the required target cell, particularly the liver, muscle and eye. The inserted DNA is incorporated into the specific body cell, where it generates the required protein encoded by the therapeutic gene.

AAV1 delivery of the correct gene into muscle cells is the basis for the successful treatment of a rare disease called familial lipoprotein lipase deficiency.

Familial Lipoprotein Lipase Deficiency

The *lipoprotein lipase* (*LPL*) gene on human chromosome 8p normally encodes an enzyme that aids in the breakdown of fats in blood.

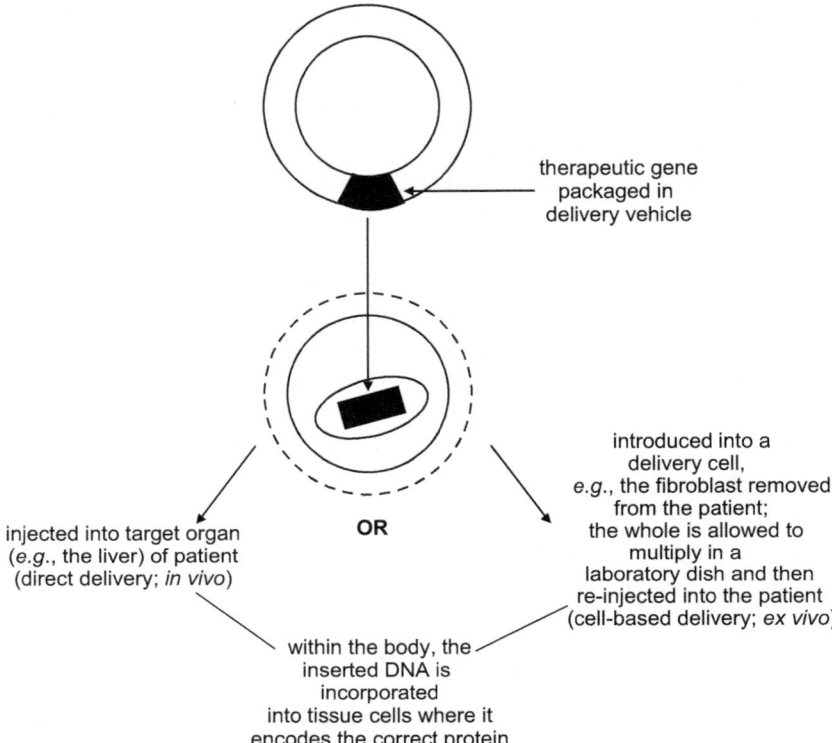

therapeutic gene
packaged in
delivery vehicle

injected into target organ
(*e.g.*, the liver) of patient
(direct delivery; *in vivo*)

OR

introduced into a
delivery cell,
e.g., the fibroblast removed
from the patient;
the whole is allowed to
multiply in a
laboratory dish and then
re-injected into the patient
(cell-based delivery; *ex vivo*)

within the body, the
inserted DNA is
incorporated
into tissue cells where it
encodes the correct protein

Figure 8.7   Direct delivery and isolated cell-based delivery techniques.

The most common mutation (G188E) causes reduced or zero enzyme activity and leads to the rare, inherited autosomal recessive familial lipoprotein lipase deficiency, which is associated with inflammation of the pancreas (and severe attacks of pancreatitis) and enlargement of the liver and spleen. Glybera is the first gene therapy treatment approved (by the EMA) for patients with the deficiency (but detectable levels of enzyme in their blood) using AAV1 delivery of the correct gene into muscle cells, thus correcting lipoprotein lipase deficiency by enabling muscle cells to produce amounts of the enzyme. Unfortunately, the price is very high so this has offset the apparent medical success.

There has been some progress in treating hemophilia in dogs using the AAV vector to deliver the gene for the missing factor to liver cells. Most of the treated dogs had fewer bleeding incidents.

Hemophilia in animals is discussed in X-linked recessive disorders in Section 6.3.6 in Chapter 6.

8.4.7   Indirect Delivery (*Ex Vivo*)

In the *ex vivo* approach, genetically defective cells are removed from the tissue of interest in the person and cultured *in vitro*, *i.e.*, outside the cell. The therapeutic gene (as described in the *in vivo* section) is mixed with the defective target cells isolated from the patient and the cells grown in culture. The genetically modified cells thus produced are finally returned to the patient (Figure 8.7), where they produce the required protein encoded by the therapeutic gene. A wide range of delivery systems (*e.g.*, electroporation or vectors) is possible. The *ex vivo* method is used more than the *in vivo* approach and is less likely to trigger an immune response because it does not place a virus into the recipient. The method can only be applied to tissues that can survive outside the body and can be returned to the target tissues, *e.g.*, red blood cells and skin cells. The technique is well illustrated in the treatment of several severe forms of primary immune deficiency disease by hematopoietic (blood-forming) stem cell transplantation (HSCT). The patient's own bone marrow is removed and stem cells from it exposed to a vector carrying the normal gene. After some manipulations, it is intravenously injected back into the patient, where the stem cells start to grow and make normal lymphocytes that can produce the normal enzyme. Three severe immune deficiency diseases have been treated in this way, which will now be discussed.

Adenosine Deaminase Deficiency

The *adenosine deaminase* (*ADA*) gene on human chromosome 20q encodes the enzyme adenosine deaminase. This occurs in all cells, especially lymphocytes (white blood cells), which together with lymphoid tissues comprise the immune system. The enzyme eliminates deoxyadenosine, which is toxic to lymphocytes. More than 70 mutations in the *ADA* gene, usually a single amino acid change, result in an unstable or missing enzyme. This leads to dead or greatly reduced lymphocytes and the loss of infection-fighting ability. This is a very rare form of severe combined immune deficiency (SCID), termed ADA-SCID, and children with the disease rarely live beyond two years. It was the first hereditary disease to be successfully treated by gene therapy and in a recent trial 23 babies out of 23 babies have been successfully treated.

The first multinational drug company (GSK) has been granted European commission approval to implement gene therapy. In this case, Strimvelis has been marketed for the treatment of ADA-SCID by HSCT. This will encourage such companies to develop gene therapy, already tackled by smaller companies, to treat more common diseases, such as sickle cell anemia, *etc.*, although the high price remains a serious consideration (see Glybera in Section 8.4.6).

X-linked Severe Combined Immune Deficiency (XSCID)

The *interleukin-2 receptor gamma* (*IL2RG*) gene on the human Xq chromosome encodes the common gamma chain protein critical for normal immune system function. More than 300 mutations have been associated with the X-linked recessive disease, attended with loss of the immune system, recurring chest infections unresponsive to antibiotics and probable death within two years. It is the most common form of SCID and affects about 1 in 50 000 newborns. XSCID was one of the first genetic disorders treated by gene therapy. Very unfortunately, leukemia developed in some patients. It was found that the earlier viral vectors used included DNA sequences that enhanced integrated oncogenes. This failure heralded a massive retreat from gene therapy until recently. Newer viral vectors are now being used and early results indicate that they are effective and safe. Success has also been reported for gene therapy treatment of a rare immune-deficient disease called Wiskott–Aldrich (W–A) syndrome.

Wiskott–Aldrich (W–A) Syndrome

The relevant *Wiskott–Aldrich syndrome* (*WAS*) gene is located on the Xp human chromosome. The gene encodes the WAS protein, which assists in the interaction of white blood cells with foreign invaders, which is essential for the protection of the body from infection. W–A syndrome arises from one of a reported 350 mutations in the *WAS* gene, leading to a shortened and ineffective or absent WAS protein. Children with the mutation have a reduced ability to fight infection and show bleeding episodes, skin infections and autoimmune diseases. There is a short life expectancy, 20 years or so, and often prolonged spells in hospital. The very rare disease seldom occurs in females. In an admittedly limited trial in the US, seven children with W–A syndrome were treated by gene therapy. In the two years since the treatment, most of the symptoms have been resolved, including restoration of immune function and no visits to the hospital—so far, an altogether impressive outcome. Possible gene therapy treatment using CRISPR removal of the faulty gene is outlined in 8.4.9.

8.4.8 Non-viral Vector Gene Therapy

Although viral vectors are very good at entering cells (sometimes specific types), they can carry a limited amount of DNA and may cause an immune response. These limitations may be overcome by using a non-viral vector, such as a plasmid (Section 2.2.1 in Chapter 2). This may be wrapped inside a liposome that can fuse with cell membranes. Plasmids and liposomes are, however, much less efficient and less specific than viral vectors at gene transfer to a cell, but they can carry larger genes and are unlikely to initiate an immune response.

8.4.9 Gene Therapy using CRISPR/Cas9

An enormous addition to the gene therapy armory of the geneticist is the recent use of CRISPR/Cas9 genome editing in living cells. Several genome-engineering approaches have been used, *e.g.*, zinc-finger nucleases (ZFNs) and transcription activator-like effector nucleases (TALENs), but CRISPR has caught the imagination for its simplicity and adaptability, and this approach has been mainly emphasized in the book. Over a 1000 research reports were published in 2015 in which the technique, or variants, feature. This and the wide areas of genetics addressed are testament to the strong interest the method has invoked. It is also worth pointing out that the birth of the technique in 2012 arose from somewhat esoteric research on bacterial protection against foreign viruses, in which CRISPR is involved.

CRISPR/Cas9 System

CRISPR (clustered regularly interspaced short palindromic repeat!) can modify precisely the DNA or its expression *at a specific location* in the genome of an animal.

- An approximately 20 base RNA segment (protospacer element or spacer), synthesized to match a portion of a specific target DNA sequence, is a type of homing device. It is made from the terminus of a short synthetic RNA composed of a "scaffold" sequence for Cas9 binding. The two components are termed a single guide or guide RNA (gRNA).
- The gRNA binds, activates and guides an endonuclease enzyme Cas9 (CRISPR associated protein, usually derived from *Streptococcus pyogenes*—a bacterium that causes strep throat) to the target DNA site defined by the spacer and which is opened up (Figure 8.8).
- The target DNA must be adjacent to an essential, specific three nucleotide protospacer adjacent motif (PAM) on the non-targeted DNA strand. Provided also that there is sufficient homology between the spacer and the target DNA, the enzyme will effect a double-stranded break three to four nucleotides upstream of the PAM.
- The double-strand breaks (DSB) must be repaired. A cell tries by using non-homologous end joining (NHEJ) and homology directed repair (HDR; Section 4.6 in Chapter 4). NHEJ introduces small INDEL mutations (Section 4.3.4 in Chapter 4), which ideally knock-out the gene by frameshift mutations or premature stop codons. Endogenous HDR, which is error-free, can be exploited to effect gene editing. The break may be rapidly repaired by adding a donor template with a desired modified DNA sequence to gRNA and Cas9, which may, for example, obviate a particular mutation (knock-in). The repair template must also contain 50–80 bp DNA sequences, which will bind by homology upstream and downstream of the blunt ends of the cleaved target.
- Since its inception, a number of modifications of the wild-type Cas9 have been made to make it more specific to the target and avoid the "off-targeting" editing that Cas9 is prone to, which might alter other genes (very undesirable).

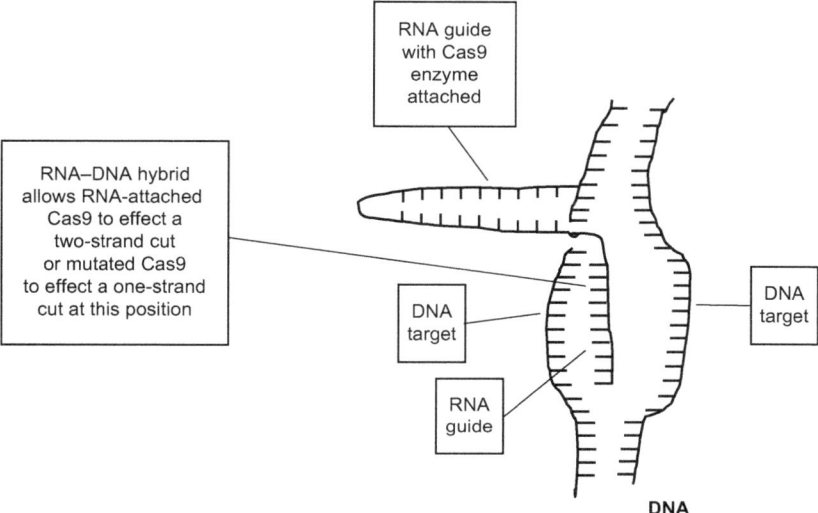

Figure 8.8 The almost finished product in CRISPR/Cas9 editing. gRNA has two linked components. CRISPR RNA (crRNA), which contains the sequence complementary to the DNA target, and an RNA segment that base pairs with the second component: trans-activating CRISPR RNA (tracrRNA), which is required for Cas9-mediated target cleavage. gRNA attached to the Cas9 enzyme recognizes a *ca.* 20 base target site adjacent (5′ end) to an essential protospacer adjacent motif (PAM). PAM is a short, three nucleotide base sequence, 5′-NGG-3′. A double-stranded cut is made three to four nucleotides upstream of PAM.

- Cas9 nickase is a mutated version, which effects a single break (nick) of the binding DNA strand. It can generate a DSB, consisting of two slightly offsetting single breaks on opposite DNA strands using two separate sgCas9s. Off-target breaks are reduced.
- dCas9 is a catalytically inactive form. gRNA can still guide dCas9 to a chosen genomic area, but the DNA is not cleaved. Fusion of dCas9 with activator or repressor peptides can be used for gene activation or silencing. Fused to a fluorescent protein, dCas9 can locate and show up a specific gene region with a known DNA sequence in the presence of many genes.
- Gene therapy using CRISPR is more likely to focus on monogenic diseases (Figure 1.13) in the first instance, although the ability of CRISPR to edit genes in several places simultaneously suggests that the much more difficult

multigenic diseases will be tackled in time. Some examples of very many types of applications are:

○ The feasibility of treating X-linked retinitis pigmentosa using CRISPR has been demonstrated (Section 6.3.5 in Chapter 6).

○ The muscle function of a mouse model for DMD has been improved by CRISPR snipping out a faulty exon. The vector AAV was used to carry DNA encoding gRNA and Cas9 into every muscle of the mouse and cut out the faulty exon. Muscle function was improved throughout the body, including the heart.

○ Using HDR, CRISPR has repaired a single base mutation causing cataracts in mice.

• Both the *in vivo* (Section 8.4.6) and *ex vivo* (Section 8.4.7) approaches have been used in a number of different animals. UK scientists have now been granted license to carry out gene modification of embryos donated from IVF treatment using the CRISPR/Cas9 system. The object is to further understand the genes that humans need at an early stage to develop successfully and the reasons that some women have miscarriages. The embryos would be destroyed after no more than 14 days.

These are just inklings of the possibilities for this mode of treatment for diseases with the strong proviso that shortcomings in the method (such as safe and effective transfer to the cell—problems that apply to any gene therapy mode) have to be addressed before it becomes a clinical possibility.

8.4.10   Human Tissue-engineered Therapy

Tissue engineering is the regeneration of biological tissue through the use of cells and supporting biomolecules. The first tissue-engineered product based on stem-cell therapy has been approved by the EMA. It will be used for the treatment of limbal stem-cell deficiency. In some cases of damage (diseases and burns) to the eye in the border of the cornea (clear front part of the eye) with the sclera (white part of the eye), limbal stem cells, which keep the cornea clear and healthy, are not completely replaced naturally. This leads to as many as 20 eye problems and sometimes eventual blindness. In the treatment, the patient's own limbal stem cells from a small part of any healthy limbal tissue (cornea) are removed. These are cultured *ex vivo* and used to make Holoclar by growing them on a fibrin membrane shaped like a contact lens. The membrane and cells are then implanted into the eye after removal of the damaged area. Once implanted, the corneal cells in Holoclar help replace the corneal surface, while the limbal stem cells are a reservoir for new corneal cells. The procedure has been generally successful in trials and may soon become available. Since the stem cells used originate from the person's own body, rejection is unlikely.

8.5   GENE-DIRECTED DRUGS

Gene therapy is probably the ultimate remedy for correcting undesired traits, particularly diseases. There are, however, substantial difficulties in its implementation. The next best approach is to tackle the actual site of the gene mutation that results in the disease with drugs rather than just correcting the symptoms of the disease. A few examples in this section are representative of the approach. A number are in clinical trials. A few have been approved by US and/or European agencies and have proved effective in a limited number of persons who have the relevant rare disease. A serious problem is the very high cost of the various drugs for a very rare population. This may lessen when less rare diseases are tackled, aided by the information already gleaned.

Cystic Fibrosis (CF)

The CFTR protein allows transport of chloride ions across the cell membrane. Many types of mutation, in various ways, prevent the normal protein from carrying out this task (Section 4.3.7 in Chapter 4). Gene therapy trials are in an advanced state, but specific drugs are also being designed to target the restoration of the *trans*-epithelial ion transport corrupted by these mutations. Three classes of mutation have so far been addressed, see Figure 4.1.

1. **Class 3:** The underlying G551D–CFTR nonsense mutation (present in 4% of patients with cystic fibrosis) causes the defective channel-gating in response to cellular signals.

Ivacaftor (Kalydeco) is a simple molecule used as a daily medicine that received FDA approval in 2012, as well as approval from European agencies. It binds to the channel pore and helps the faulty CFTR protein to function more normally once it reaches the cell surface by helping to keep the gate open. The drug increases chloride ion secretion and reduces excessive sodium ion and fluid absorption, thus thinning the heavy mucus characteristic of cystic fibrosis. The approval is for children only and with one of 10 mutations, including G551D.

2. **Class 2**: The DelF508 mutation (present in 60% of cystic fibrosis patients) disturbs protein folding and impairs delivery of CFTR to the cell surface. Lumacaftor helps the protein fold properly, enabling more of the CFTR protein to move to its position at the cell membrane. Clinical tests on over a 1000 sufferers with the DelF508 mutation have been completed using a combination of ivacaftor and lumacaftor (termed Orkambi), which would correct the faulty protein once it is moved to its correct position on the cell membrane and help it to remain open. Very encouraging results have been obtained. Better lung functioning, fewer lung infections and reduced CF-related hospital visits were reported. Orkambi (but not lumacaftor alone) has FDA approval for use with persons >12 years of age and approval is anticipated for younger sufferers, but the very high cost may prevent its general use at present.

3. **Class 1**: It has been found that a series of short molecules allow, by an unknown mechanism, the "read-through" of the stop codon resulting from a nonsense mutation. It has been suggested that one of these, ataluren (Translarna), may be used to circumvent the premature stop signal from a nonsense mutation (G542stop) associated with cystic fibrosis, which leads to the complete absence of the normal protein. Promising results have been reported in phase III trials.

Duchenne Muscular Dystrophy (DMD)

This is the most common form of a group of muscular dystrophies marked by wasting of certain muscles. It is usually restricted to the pelvic and back regions of boys. There are a large number and several types of mutation (Section 6.3.5 in Chapter 6). About 10–15% of sufferers have a single point nonsense mutation and the resulting premature stop signal results in the ribosome stopping along the mRNA (Section 3.1.2 in Chapter 3), no full dystrophin production and their having severe DMD.

Ataluren (Translarna) has obtained conditional approval from European agencies on the basis of incomplete clinical trials to treat this form of DMD. It is believed to interact with the ribosome by an unknown mechanism and allow it to read through the premature stop signal and produce the normal protein. In a milder form of the disease (Becker disease), the ribosome can skip past a deleted exon, resulting in a truncated form of the dystrophin protein.

Antisense Oligonucleotide Drugs

Synthetic oligonucleotide drugs are chemical modifications of the natural oligonucleotides (Section 1.2.8 in Chapter 1). This may involve using a phosphorothioate bond to join ribose rings, modifying the ribose ring (Figure 8.9) or even replacing the ribose ring with a morpholine group. In all cases, the usual bases are retained. The changes enhance a number of desirable properties (*e.g.*, stability to degradation by enzymes, reduced RNA toxicity and stronger binding to RNA) compared with the endogenous oligonucleotide counterparts. They carry out the same function, *i.e.*, target pre-mRNA or mRNA.

Their use is illustrated in the treatment of Duchenne muscular dystrophy (Section 6.3.5 in Chapter 6) by drisapersen (rejected at present by the FDA) and a similar acting drug, eteplirsen (Exondys, approved conditionally by the FDA), which have different structures. One mutation in DMD leads to deletion of exon 50 in pre-mRNA in the dystrophin gene. Splicing of pre-mRNA leads to linking of exon 49 to 51 and a disrupted reading frame results (Section 4.3.6 in Chapter 4). Further exon assembly to the end (exon 79) is aborted and no functional protein results. The two antisense oligonucleotide drugs drisapersen and eteplirsen are 21 and 30 nucleotides in length, respectively, and tailored to hybridize to a complementary RNA segment in exon 51 and silence it (termed "exon skipping"). This allows exon 48 to be spliced to exon 52 to restore the reading frame and for splicing to

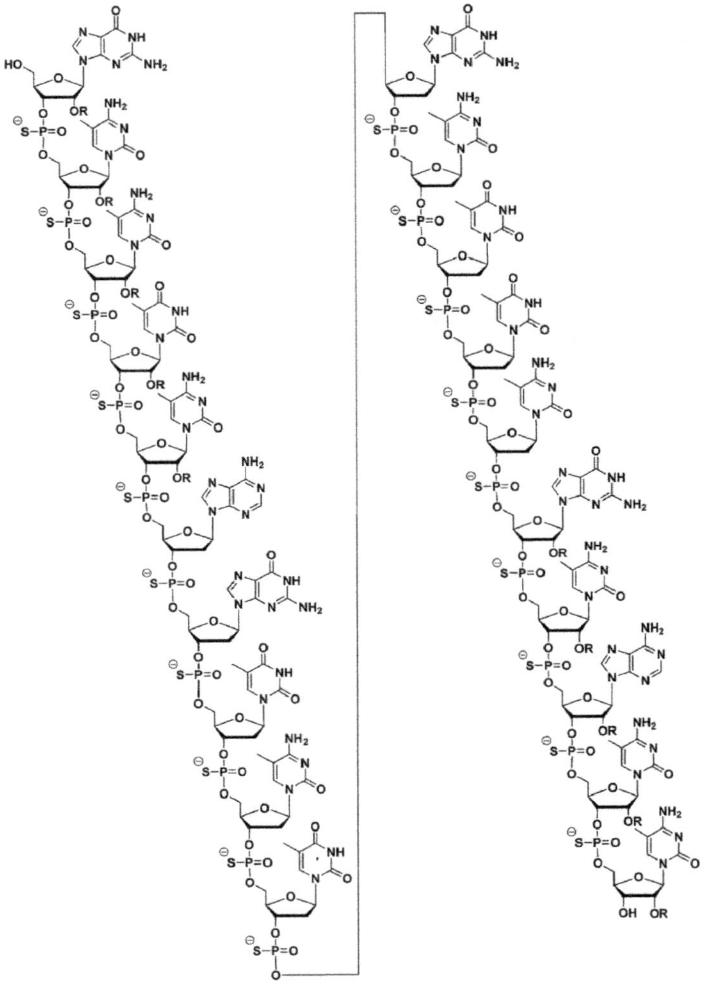

Figure 8.9 Structure of mipomersen: 5′–GCCTCAGTCTGCTTCGCACC–3′. A phosphorothioate group joins ribose sugars throughout. The central domain of 10 contiguous bases supports the degradation of the target mRNA. This is flanked by two 5-base segments and substituted ribose that strengthen the binding to the mRNA and protect the antisense oligonucleotide from enzyme cleavage. Cytosine is methylated since cytosine is more easily recognized by an innate immune system. R = OCH$_2$CH$_2$OCH$_3$.

Adapted from V. K. Sharma, R. K. Sharma and S. K. Singh, *Med. Chem. Commun.*, 2014, **5**, 1454 with permission from The Royal Society of Chemistry.

resume to the end. The result is a shorter than normal protein (material from exons 50 and 51 is missing), which is still partly functional. A milder form of the disease akin to Becker disease might result. It is estimated that 13% of boys with DMD may benefit from this treatment and have a milder prognosis and increased life span. Obviously, deletion of other exon mutations causing DMD will require different oligonucleotide drugs. These antisense oligonucleotide drugs are complex mixtures of thousands of stereoisomers arising from the many chiral centers. It is hoped that stereo-pure nucleic acids, the preparation of which is formidable, will be more effective than the racemate mixture.

The drug mipomersen (Kynamro), an antisense 20 base oligonucleotide (Figure 8.9), has FDA approval to treat homozygous familial hypercholesterolemia (Section 6.2.2 in Chapter 6). This time, it is the offending portion of the mature mRNA and not the pre-mRNA that is targeted.

Familial Hypercholesterolemia

One of the genes that, when mutated, causes familial hypercholesterolemia is the *apolipoprotein B* (*APOB*) gene on human chromosome 2p, which encodes apolipoprotein B100 (ApoB100). This protein, which is produced in the liver, can carry cholesterol in a variety of forms: high density lipoprotein (HDL) being the good form and low density lipoprotein (LDL) being the bad or atherogenic form (tending to form fatty plaques in the arteries), the latter of which is implicated in heart disease. Mutations in the gene cause altered proteins, which are less able to remove LDLs from the blood so that cholesterol levels in the blood become very high, leading to *familial defective apolipoprotein B*, a form of familial hypercholesterolemia. Possession of two copies of the altered gene leads to an increased risk of heart disease, which may appear in childhood.

Mipomersen (Kynamro) is a weekly subcutaneous injection for inhibiting the synthesis of ApoB100. It does this by forming a duplex by base pairing with a sequence within the coding region of Apo B mRNA at exon 22. The duplex is recognized and the target mRNA cleaved by an endogenous enzyme. This prevents translation of ApoB100 protein, leading to decreased ApoB100, lowered LDL cholesterol and some alleviation of the severity of the disease. It is used as an adjunct to lipid-lowering medications and diet.

8.5.1 Cancers

Most work with gene-directed drugs has concentrated on the treatment of cancer. In simple terms, cancer is a condition in which abnormal cells divide uncontrollably (the cell cycle goes amok). These abnormal cells lack the checks and balances in the normal cell cycle (Section 3.2.1 in Chapter 3) and multiply more rapidly than normal, healthy cells. They can invade nearby tissues and spread through the bloodstream and lymphatic system to other parts of the body. Genes whose mutant forms may cause normal cells to become cancerous are termed oncogenes. There are many types of cancer present in all parts of the body. The FDA has approved therapies for the treatment of cancers of at least 25 different organs. Many more are currently undergoing preclinical testing on animals. Chemotherapy drugs are cytotoxic and act against all actively dividing cells, killing normal and cancerous cells. In contrast, cancer-targeted therapy drugs (cytostatic) interfere with specific chemicals that are involved in cancer cell growth and survival, and therefore block tumor-cell proliferation. The approach to the use of cancer-targeted drugs, which prevent uncontrolled proliferation of tumor cells, can be illustrated by reference to two particular cancers: chronic myeloid leukemia and melanoma. These drugs are enzyme inhibitors that selectively bind at ATP sites, which are intimately involved in the mechanism of action of the mutant-damaging enzyme and thereby inhibit its action. The selectivity is the key to their success since interference in the ATP participations elsewhere in the body would be intolerable.

Chronic Myeloid Leukemia

Leukemia results from an increased number of certain types of white blood cell. In chronic myeloid leukemia, the type of blood cell that proliferates abnormally originates in blood-forming (myeloid) tissue of the bone marrow.

The BCR-ABL tyrosine kinase (Figure 4.3) transfers a phosphate group from ATP to tyrosine residues on various substrates (signaling molecules) and causes excess tyrosine kinase with a concomitant proliferation of myeloid cells. One of the early successes of gene-directed drugs was the use of imatinib (Gleevec). This blocks the binding of ATP to the BCR-ABL complex and inhibits the enzyme's excessive kinase activity. It has been found that about 85% of patients benefit from the drug. Second generation tyrosine kinase inhibitors, analogs of imatinib, are being used to overcome some mutations that can cause resistance to imatinib.

Melanoma

The *v-RAF murine sarcoma viral oncogene homolog B (BRAF)* gene is a member of the oncogene class that, when mutated, has the potential to cause normal cells to become cancerous cells. The *BRAF* gene on chromosome 7q encodes BRAF kinase. The enzyme is a necessary participant in the essential RAS/RAF signaling pathway, which regulates cell growth and division, differentiation and apoptosis of the cell. A prominent missense mutation, V600E, on *BRAF* encodes an overactive oncogene-mutated BRAF protein, which causes continual signaling and stimulates abnormal cells to divide uncontrollably and give rise to melanoma, an aggressive form of skin cancer. It is not inherited and is acquired in only certain cell types. Various mutations of the gene give rise to a number of disorders. Vemurafenib (Zelboraf) is effective only for the treatment of the V600E-mutated variety of melanoma, which is found, however, in 50% of melanomas. It selectively binds to an ATP binding site in the V600E mutant kinase and inhibits its action. This renders the encoded oncogenic protein fully inactive and eradicates cancer cells. Downstream proliferating signaling is inhibited and desirable apoptosis results. Targeting the DNA repair mechanisms used in cancer cells can lead to blocked tumor growth.

Ovarian Cancer

As well as breast cancer (Section 7.2.5 in Chapter 7), *BRCA1* and *2* genes are implicated in ovarian cancer and account for 10–20% of all ovarian cancers.

Poly (ADP-ribose) polymerase (PARP) are a family of enzymes with a number of functions, which include mediating base excision repair (Section 4.6 in Chapter 4) of one-strand DNA breaks. PARP enzymes help repair damaged DNA in cells by one pathway and BRCA proteins use another pathway. Cancers related to *BRCA* gene mutations appear to depend on PARP to repair damaged DNA in cancer cells and, as a consequence, to maintain abnormal cell division. Olaparib (Lynparza) is an orally administered drug used for patients with previously treated advanced ovarian cancer arising from *BRCA* mutations. As an inhibitor of the PARP pathway, it is difficult for tumor cells with a *BRCA* gene to repair damaged cells, and therefore growth of the tumor cell is slowed or terminated, leading to tumor cell death. Olaparib is also in phase III clinical trials for the treatment of breast and other cancers.

8.5.2 Epigenetic-based Drugs

An oncogene normally directs cell growth, but when it is mutated or over-expressed, it typically increases cell division and inhibits cell death, causing normal cells to become cancerous. These actions are offset by tumor suppressor genes, which retard cell proliferation. Any action within the cell that interferes with either or both of these genes is likely to promote the cancer condition. This is where epigenetic factors come into play. If the DNA on CpG islands in the promoter region of a tumor suppressor gene that is normally unmethylated becomes *excessively* methylated, the expression of the suppressor gene will be silenced—a very undesirable happening since the cancer condition is likely to be promoted. In the same way, a *low* level of acetylation of DNA histone(s) will also turn off the tumor suppressor gene. Both abnormalities may invariably lead to cell proliferation and cancer. Drugs that can inhibit by binding directly to the active site of the enzymes that are responsible for the abnormal epigenetic markings (Section 7.4) will increase desirable gene expression, switch the suppressor gene on again and likely keep the cancer cells under control. Four drugs have FDA approval for clinical use with cancer patients: two which take care of the excessive methylation focusing on *DNA* *m*ethyl*t*ransferase (DNMT), and two which promote acetylation focusing on *h*istone *deac*etylase (HDAC).

- 5-Azacitidine (Vidaza) and decitabine (Dacogen) are hypo-methylating agents with antineoplastic activity (inhibiting or preventing the growth and spread of malignant cells). At present, they are confined to treating several hematological (blood and bone marrow) cancers.
- Vorinostat, also termed SAHA (Zolinza), and romidepsin (Istodax), a bacterial bicyclic depsipeptide, are inhibitors of certain types of HDAC enzymes (there are a number of classes). The inhibition by vorinostat is particularly interesting. The drug binds to the catalytic site of the HDACs and this allows the hydroxamic portion to form a chelate complex with a nearby Zn ion in the catalytic pocket and inhibit deacetylation. These drugs prevent HDAC enzymes from removing acetyl groups from histones and help them to carry many acetyl groups and restore the normal acetylation pattern. This suppresses tumor growth and progression, induces apoptosis and encourages tumor-cell death. HDAC inhibitors have long been used for the treatment of neurological conditions and have only recently been directed to cancer therapy. Both drugs are used to treat skin problems arising from T-cell lymphomas when other treatments have failed. A number of other drugs have been approved, which use HDAC and DNMT in anti-cancer action.

Some Thoughts from the Author

I hope that I have conveyed to the reader the excitement of present day genetics. Building on the rules established by previous researchers, we are beginning to understand the wondrous transformation of a single cell into an adult animal and use the information to cure, not just alleviate, human sufferings. We must use the word beginning because many obstacles must be circumvented before gene therapy becomes an everyday reality, but become it will. Genetics has a role to play in many facets of human endeavour, only some of which have been outlined in the book. I would dearly love to be around 50–100 years from now to see the impact of genetics on our everyday experiences.

Helpful Reading Material

The following books have been very helpful in writing this book:

T. A. Brown, *Gene Cloning and DNA Analysis. An Introduction*, Wiley-Blackwell, 6th edn, 2010.

T. Brown, *Introduction to Genetics*, Garland Science, New York, 2012.

N. Carey, *The Epigenetics Revolution*, Icon Books Ltd, London, 2012.

J. Challoner, *The Cell*, Ivy Press, Lewes, UK, 2015.

S. L. Elrod and W. D. Stansfield, *Genetics,* Schaum's outlines, McGraw-Hill, 5th edn, 2010.

R. V. Knowles, *The Wonder of Genetics*, Prometheus Books, New York, 2010.

E. Passarge, *Color Atlas of Genetics*, Thieme, Stuttgart, New York, 4th edn, 2013.

M. Ridley, *Genome, The Autobiography of a Species in 23 Chapters*, Harper Collins, New York, 2000.

T. R. Robinson, *Genetics for Dummies*, Wiley, Hoboken NJ, 2nd edn, 2010.

C. P. Schaaf, J. Zschocke and L. Potocki, *Human Genetics: From Molecules to Medicine*, Lippincott Williams and Wilkins, Baltimore, 2012.

Animal Genetics for Chemists
By Ralph G. Wilkins
© Ralph G. Wilkins 2017
Published by the Royal Society of Chemistry, www.rsc.org

A "Very Short Introduction" series published by Oxford University Press, New York:
 T. Allen and G. Cowling, *The Cell*, 2011.
 L. Wolpert, *Developmental Biology*, 2011.
 J. Stack, *Stem Cells*, 2013.
 J. Stack, *Genes*, 2014.

HELPFUL REFERENCE MATERIAL

R. C. King, W. D. Stansfield and P. K. Mulligan, *A Dictionary of Genetics*, OUP, New York, 7th edn, 2006.

Oxford Dictionary of Biology, OUP, Oxford, 6th edn, 2008.

The Bantam Medical Dictionary, Random House, New York, 6th edn, 2009.

The Merck Index[†]*: An Encyclopedia of Chemicals, Drugs and Biologicals*, Royal Society of Chemistry, Cambridge, 15th edn, 2013.

Genetics Home Reference, a free online service of the National Library of Medicine, part of the National Institutes of Health. It is an excellent source of information on human diseases with a genetic component.

[†]The name THE MERCK INDEX is owned by Merck Sharp & Dohme Corp., a subsidiary of Merck & Co., Inc., Whitehouse Station, N.J., U.S.A., and is licensed to The Royal Society of Chemistry for use in the U.S.A. and Canada.

Subject Index